TOXICITY OF MILITARY SMOKES AND OBSCURANTS

VOLUME 2

SUBCOMMITTEE ON MILITARY SMOKES AND OBSCURANTS

COMMITTEE ON TOXICOLOGY

BOARD ON ENVIRONMENTAL STUDIES AND TOXICOLOGY

COMMISSION ON LIFE SCIENCES

NATIONAL RESEARCH COUNCIL

NATIONAL ACADEMY PRESS
WASHINGTON, D.C.

NATIONAL ACADEMY PRESS   2101 Constitution Ave., N.W.   Washington, D.C. 20418

NOTICE: The project that is the subject of this report was approved by the Governing Board of the National Research Council, whose members are drawn from the councils of the National Academy of Sciences, the National Academy of Engineering, and the Institute of Medicine. The members of the committee responsible for the report were chosen for their special competencies and with regard for appropriate balance.

The National Academy of Sciences is a private, nonprofit, self-perpetuating society of distinguished scholars engaged in scientific and engineering research, dedicated to the furtherance of science and technology and to their use for the general welfare. Upon the authority of the charter granted to it by the Congress in 1863, the Academy has a mandate that requires it to advise the federal government on scientific and technical matters. Dr. Bruce Alberts is president of the National Academy of Sciences.

The National Academy of Engineering was established in 1964, under the charter of the National Academy of Sciences, as a parallel organization of outstanding engineers. It is autonomous in its administration and in the selection of its members, sharing with the National Academy of Sciences the responsibility for advising the federal government. The National Academy of Engineering also sponsors engineering programs aimed at meeting national needs, encourages education and research, and recognizes the superior achievements of engineers. Dr. William A. Wulf is president of the National Academy of Engineering.

The Institute of Medicine was established in 1970 by the National Academy of Sciences to secure the services of eminent members of appropriate professions in the examination of policy matters pertaining to the health of the public. The Institute acts under the responsibility given to the National Academy of Sciences by its congressional charter to be an adviser to the federal government and, upon its own initiative, to identify issues of medical care, research, and education. Dr. Kenneth I. Shine is president of the Institute of Medicine.

The National Research Council was organized by the National Academy of Sciences in 1916 to associate the broad community of science and technology with the Academy's purposes of furthering knowledge and advising the federal government. Functioning in accordance with general policies determined by the Academy, the Council has become the principal operating agency of both the National Academy of Sciences and the National Academy of Engineering in providing services to the government, the public, and the scientific and engineering communities. The Council is administered jointly by both Academies and the Institute of Medicine. Dr. Bruce M. Alberts and Dr. William A. Wulf are chairman and vice chairman, respectively, of the National Research Council.

This project was supported by Contract Nos. DAMD 17-89-C-9086 and DAMD 17-99-C-9049 between the National Academy of Sciences and the U.S. Department of Defense. Any opinions, findings, conclusions, or recommendations expressed in this publication are those of the author(s) and do not necessarily reflect the view of the organizations or agencies that provided support for this project.

International Standard Book Number 0-309-06329-9

Additional copies of this report are available from:

National Academy Press
2101 Constitution Ave., NW
Box 285
Washington, DC 20055
800-624-6242
202-334-3313 (in the Washington metropolitan area)
http://www.nap.edu

Copyright 1999 by the National Academy of Sciences. All rights reserved.

SUBCOMMITTEE ON MILITARY SMOKES AND OBSCURANTS

MICHELE A. MEDINSKY *(Chair)*, Chemical Industry Institute of Toxicology, Research Triangle Park, North Carolina
KEVIN E. DRISCOLL, The Procter and Gamble Company, Cincinnati, Ohio
CHARLES E. FEIGLEY, University of South Carolina School of Public Health, Columbia, South Carolina
DONALD E. GARDNER, Inhalation Toxicology Associates, Raleigh, North Carolina
SIDNEY GREEN, Howard University, Washington, D.C.
ROGENE F. HENDERSON, Lovelace Respiratory Research Institute, Albuquerque, New Mexico
CAROLE A. KIMMEL, U.S. Environmental Protection Agency, Washington, D.C.
HANSPETER R. WITSCHI, University of California, Davis, California
GAROLD S. YOST, University of Utah, Salt Lake City, Utah

Staff

KULBIR S. BAKSHI, Program Director for the Committee on Toxicology
ABIGAIL STACK, Project Director
RUTH E. CROSSGROVE, Editor
LINDA V. LEONARD, Senior Project Assistant
LUCY V. FUSCO, Project Assistant
CHRISTINE PHILLIPS, Project Assistant

Sponsor

U.S. DEPARTMENT OF DEFENSE

COMMITTEE ON TOXICOLOGY

BAILUS WALKER, JR. *(Chair)*, Howard University Medical Center and American Public Health Association, Washington, D.C.
MELVIN E. ANDERSEN, Colorado State University, Denver, Colorado
GERMAINE M. BUCK, State University of New York at Buffalo, Buffalo, New York
GARY P. CARLSON, Purdue University, West Lafayette, Indiana
JACK H. DEAN, Sanofi Pharmaceuticals, Inc., Malverne, Pennsylvania
ROBERT E. FORSTER II, University of Pennsylvania, Philadelphia, Pennsylvania
PAUL M.D. FOSTER, Chemical Industry Institute of Toxicology, Research Triangle Park, North Carolina
DAVID W. GAYLOR, U.S. Food and Drug Administration, Jefferson, Arkansas
JUDITH A. GRAHAM, U.S. Environmental Protection Agency, Research Triangle Park, North Carolina
SIDNEY GREEN, Howard University, Washington, D.C.
WILLIAM E. HALPERIN, National Institute for Occupational Safety and Health, Cincinnati, Ohio
CHARLES H. HOBBS, Lovelace Respiratory Research Institute and Lovelace Biomedical and Environmental Research Institute, Albuquerque, New Mexico
FLORENCE K. KINOSHITA, Hercules Incorporated, Wilmington, Delaware
MICHAEL J. KOSNETT, University of Colorado Health Sciences Center, Denver, Colorado
MORTON LIPPMANN, New York University School of Medicine, Tuxedo, New York
THOMAS E. MCKONE, Lawrence Berkeley National Laboratory and University of California, Berkeley, California
ERNEST E. MCCONNELL, ToxPath, Inc., Raleigh, North Carolina
DAVID H. MOORE, Battelle Memorial Institute, Bel Air, Maryland
GÜNTER OBERDÖRSTER, University of Rochester, Rochester, New York
JOHN L. O'DONOGHUE, Eastman Kodak Company, Rochester, New York
GEORGE M. RUSCH, AlliedSignal, Inc., Morristown, New Jersey
MARY E. VORE, University of Kentucky, Lexington, Kentucky
ANNETTA P. WATSON, Oak Ridge National Laboratory, Oak Ridge, Tennessee

Staff

KULBIR S. BAKSHI, Program Director
SUSAN N.J. PANG, Program Officer
ABIGAIL STACK, Program Officer
RUTH E. CROSSGROVE, Publications Manager
KATHERINE IVERSON, Manager of the Toxicology Information Center
CATHERINE M. KUBIK, Senior Program Assistant
LINDA V. LEONARD, Senior Project Assistant
LUCY V. FUSCO, Project Assistant
CHRISTINE PHILLIPS, Project Assistant

BOARD ON ENVIRONMENTAL STUDIES AND TOXICOLOGY

GORDON ORIANS *(Chair)*, University of Washington, Seattle, Washington
DONALD MATTISON *(Vice Chair)*, March of Dimes, White Plains, New York
DAVID ALLEN, University of Texas, Austin, Texas
MAY R. BERENBAUM, University of Illinois, Urbana, Illinois
EULA BINGHAM, University of Cincinnati, Cincinnati, Ohio
PAUL BUSCH, Malcolm Pirnie, Inc., White Plains, New York
PETER L. DEFUR, Virginia Commonwealth University, Richmond, Virginia
DAVID L. EATON, University of Washington, Seattle, Washington
ROBERT A. FROSCH, Harvard University, Cambridge, Massachusetts
JOHN GERHART, University of California, Berkeley, California
MARK HARWELL, University of Miami, Miami, Florida
ROGENE HENDERSON, Lovelace Respiratory Research Institute, Albuquerque, New Mexico
CAROL HENRY, American Petroleum Institute, Washington, D.C.
BARBARA HULKA, University of North Carolina, Chapel Hill, North Carolina
DANIEL KREWSKI, Health Canada and University of Ottawa, Ottawa, Ontario
JAMES A. MACMAHON, Utah State University, Logan, Utah
MARIO J. MOLINA, Massachusetts Institute of Technology, Cambridge, Massachusetts
CHARLES O'MELIA, Johns Hopkins University, Baltimore, Maryland
KIRK SMITH, University of California, Berkeley, California
MARGARET STRAND, Oppenheimer Wolff Donnelly & Bayh, LLP, Washington, D.C.
TERRY F. YOSIE, Chemical Manufacturers Association, Arlington, Virginia

Senior Staff

JAMES J. REISA, Director
DAVID J. POLICANSKY, Associate Director and Senior Program Director for Applied Ecology
CAROL A. MACZKA, Senior Program Director for Toxicology and Risk Assessment
RAYMOND A. WASSEL, Senior Program Director for Environmental Sciences and Engineering
KULBIR BAKSHI, Program Director for the Committee on Toxicology
LEE R. PAULSON, Program Director for Resource Management

COMMISSION ON LIFE SCIENCES

MICHAEL T. CLEGG *(Chair)*, University of California, Riverside, California
PAUL BERG *(Vice Chair)*, Stanford University, Stanford, California
FREDERICK R. ANDERSON, Cadwalader, Wickersham & Taft, Washington, D.C.
JOHN C. BAILAR III, University of Chicago, Chicago, Illinois
JOANNA BURGER, Rutgers University, Piscataway, New Jersey
SHARON L. DUNWOODY, University of Wisconsin, Madison, Wisconsin
DAVID EISENBERG, University of California, Los Angeles, California
JOHN EMMERSON, Portland, Oregon
NEAL FIRST, University of Wisconsin, Madison, Wisconsin
DAVID J. GALAS, Keck Graduate Institute of Applied Science, Claremont, California
DAVID V. GOEDDEL, Tularik, Inc., South San Francisco, California
ARTURO GOMEZ-POMPA, University of California, Riverside, California
COREY S. GOODMAN, University of California, Berkeley, California
HENRY HEIKKINEN, University of Northern Colorado, Greeley, Colorado
BARBARA S. HULKA, University of North Carolina, Chapel Hill, North Carolina
HANS J. KENDE, Michigan State University, East Lansing, Michigan
CYNTHIA KENYON, University of California, San Francisco, California
MARGARET G. KIDWELL, University of Arizona, Tucson, Arizona
BRUCE R. LEVIN, Emory University, Atlanta, Georgia
OLGA F. LINARES, Smithsonian Tropical Research Institute, Miami, Florida
DAVID LIVINGSTON, Dana-Farber Cancer Institute, Boston, Massachusetts
DONALD R. MATTISON, March of Dimes, White Plains, New York
ELLIOT M. MEYEROWITZ, California Institute of Technology, Pasadena, California
ROBERT T. PAINE, University of Washington, Seattle, Washington
RONALD R. SEDEROFF, North Carolina State University, Raleigh, North Carolina
ROBERT R. SOKAL, State University of New York, Stony Brook, New York
CHARLES F. STEVENS, The Salk Institute, La Jolla, California
SHIRLEY M. TILGHMAN, Princeton University, Princeton, New Jersey
JOHN L. VANDEBERG, Southwest Foundation for Biomedical Research, San Antonio, Texas
RAYMOND L. WHITE, University of Utah, Salt Lake City, Utah

WARREN R. MUIR, Executive Director

OTHER REPORTS OF
THE BOARD ON ENVIRONMENTAL STUDIES AND TOXICOLOGY

Ozone-Forming Potential of Reformulated Gasoline (1999)
Risk-Based Waste Classification in California (1999)
Arsenic in Drinking Water (1999)
Research Priorities for Airborne Particulate Matter: I. Immediate Priorities and a Long-Range Research Portfolio (1998)
Brucellosis in the Greater Yellowstone Area (1998)
The National Research Council's Committee on Toxicology: The First 50 Years (1997)
Toxicologic Assessment of the Army's Zinc Cadmium Sulfide Dispersion Tests (1997)
Carcinogens and Anticarcinogens in the Human Diet (1996)
Upstream: Salmon and Society in the Pacific Northwest (1996)
Science and the Endangered Species Act (1995)
Wetlands: Characteristics and Boundaries (1995)
Biologic Markers (5 reports, 1989-1995)
Review of EPA's Environmental Monitoring and Assessment Program (3 reports, 1994-1995)
Science and Judgment in Risk Assessment (1994)
Ranking Hazardous Waste Sites for Remedial Action (1994)
Pesticides in the Diets of Infants and Children (1993)
Issues in Risk Assessment (1993)
Setting Priorities for Land Conservation (1993)
Protecting Visibility in National Parks and Wilderness Areas (1993)
Dolphins and the Tuna Industry (1992)
Hazardous Materials on the Public Lands (1992)
Science and the National Parks (1992)
Animals as Sentinels of Environmental Health Hazards (1991)
Assessment of the U.S. Outer Continental Shelf Environmental Studies Program, Volumes I-IV (1991-1993)
Human Exposure Assessment for Airborne Pollutants (1991)
Monitoring Human Tissues for Toxic Substances (1991)
Rethinking the Ozone Problem in Urban and Regional Air Pollution (1991)
Decline of the Sea Turtles (1990)

Copies of these reports may be ordered from the National Academy Press
(800) 624-6242 or (202) 334-3313
www2.nap.edu

OTHER REPORTS OF
THE COMMITTEE ON TOXICOLOGY

Assessment of Exposure-Response Functions for Rocket-Emissions Toxicants (1998)
Review of A Screening Level Risk Assessment for the Naval Air Facility at Atsugi, Japan (Letter Report) (1998)
Toxicity of Military Smokes and Obscurants, Volume 1 (1997)
Review of Acute Human-Toxicity Estimates for Selected Chemical-Warfare Agents (1997)
The National Research Council's Committee on Toxicology: The First 50 Years (1997)
Toxicologic Assessment of the Army's Zinc Cadmium Sulfide Dispersion Tests (1997)
Toxicologic Assessment of the Army's Zinc Cadmium Sulfide Dispersion Tests: Answers to Commonly Asked Questions (1997)
Toxicity of Alternatives to Chlorofluorocarbons: HFC-134a and HCFC-123 (1996)
Permissible Exposure Levels for Selected Military Fuel Vapors (1996)
Spacecraft Maximum Allowable Concentrations for Selected Airborne Contaminants, Volume 1 (1994), Volume 2 (1996), Volume 3 (1996)
Nitrate and Nitrite in Drinking Water (1995)
Guidelines for Chemical Warfare Agents in Military Field Drinking Water (1995)
Review of the U.S. Naval Medical Research Institute's Toxicology Program (1994)
Health Effects of Permethrin-Impregnated Army Battle-Dress Uniforms (1994)
Health Effects of Ingested Fluoride (1993)
Guidelines for Developing Community Emergency Exposure Levels for Hazardous Substances (1993)
Guidelines for Developing Spacecraft Maximum Allowable Concentrations for Space Station Contaminants (1992)

Preface

THE U.S. ARMY uses smokes and obscurants to screen armed forces from view, signal friendly forces, and mark positions. Military personnel are exposed to smokes and obscurants during training exercises. The Army would like to ensure that exposures to these substances do not adversely affect the health of Army personnel or the public living and working near military-training facilities. To assist with this effort, the Army requested the National Research Council (NRC) to independently review the available toxicity data on several obscuring smokes and recommend exposure guidance levels for each. In response, the NRC's Committee on Toxicology (COT) convened the Subcommittee on Military Smokes and Obscurants, which prepared this report. The report reviews toxicity data and recommends exposure guidance levels for four obscuring smokes: white phosphorus, brass, titanium dioxide, and graphite.

Several individuals assisted the subcommittee by providing information on the uses and toxicity of the smokes addressed in this report. We gratefully acknowledge Colonel Francis L. O'Donnell, Major James Martin, Colonel David Wilder, and the Office of the Surgeon General of the U.S. Army for their interest and support of this project. We also thank Winnifred Palmer, Sandra Thomson, and Michael Burnham from the U.S. Army for providing information to the subcommittee.

This report has been reviewed in draft form by individuals chosen for their diverse perspectives and technical expertise in accordance with procedures for reviewing NRC and Institute of Medicine reports. The purpose of this independent review is to provide candid and critical comments that will assist the NRC in making the published report as sound as possible and to ensure that the report meets institutional standards for objectivity, evidence, and responsiveness to the study charge.

The review comments and draft manuscript remain confidential to protect the integrity of the deliberative process. We wish to thank the following individuals, who are neither officials nor employees of the NRC, for their participation in the review of this report: John Doull, University of Kansas Medical Center; Robert Forster, University of Pennsylvania School of Medicine; Charles Hobbs, Lovelace Respiratory Research Institute; Florence Kinoshita, Hercules Incorporated; Richard Schlessinger, New York University Medical Center; and Loren Koller, Oregon State University (Review Coordinator).

The individuals listed above have provided many constructive comments and suggestions. It must be emphasized, however, that responsibility for the final content of this report rests entirely with the authoring subcommittee and the NRC.

We are also grateful for the assistance of the NRC staff in the preparation of this report. The subcommittee wishes to acknowledge Kulbir Bakshi, program director of the Committee on Toxicology and Abigail Stack, project director for this report. Other staff members contributing to this report were Paul Gilman, former executive director of the Commission on Life Sciences; James Reisa, director of the Board on Environmental Studies and Toxicology; Carol Maczka, senior program director for toxicology and risk assessment; Ruth Crossgrove, editor; and Lucy Fusco, Linda Leonard, and Christine Phillips, project assistants.

Finally, we would like to thank all the members of the subcommittee for their expertise and dedicated effort throughout the study.

> Michele A. Medinsky, Ph.D.
> *Chair*, Subcommittee on Military Smokes and Obscurants
>
> Bailus Walker Jr., Ph.D., M.P.H.
> *Chair*, Committee on Toxicology

Contents

LIST OF ABBREVIATIONS XV

SUMMARY 1

1 INTRODUCTION 7
  The Subcommittee's Task, 9
  Definitions of Exposure Guidance Levels, 10
  Approach to Developing Exposure Guidance Levels, 13
  Organization of the Report, 15
  References, 16

2 WHITE PHOSPHORUS SMOKE 18
  Background Information, 18
  Toxicokinetics, 22
  Toxicity Summary: White Phosphorus and White Phosphorus
    Smoke, 22
  Previous Recommended Exposure Limits, 29
  Subcommittee Evaluation and Recommendations, 36
  Research Needs, 40
  References, 41

3 BRASS SMOKE 45
  Background Information, 45
  Toxicokinetics, 46
  Toxicity Summary, 47
  Previous Recommended Exposure Limits, 60

Subcommittee Evaluation and Recommendations, 61
Research Needs, 64
References, 65

4 TITANIUM DIOXIDE SMOKE 68
Background Information, 68
Toxicokinetics, 69
Toxicity Summary, 70
Previous Recommended Exposure Limits, 84
Subcommittee Evaluation and Recommendations, 84
Research Needs, 91
References, 91

5 GRAPHITE SMOKE 97
Background Information, 97
Toxicokinetics, 98
Toxicity Summary, 98
Previous Recommended Exposure Limits, 104
Subcommittee Evaluation and Recommendations, 106
Research Needs, 109
References, 110

LIST OF ABBREVIATIONS

ACGIH	American Conference of Governmental Industrial Hygienists
ATSDR	Agency for Toxic Substances and Disease Registry
BAL	bronchoalveolar lavage
COT	Committee on Toxicology
CT	the product of concentration and time
DOD	U.S. Department of Defense
EEGL	emergency exposure guidance level
EPA	U.S. Environmental Protection Agency
LC_{50}	lethal concentration for 50% of the test animals
LCt_{50}	lethal concentration multiplied by exposure time for 50% of the test animals
LD_{50}	lethal dose for 50% of the test animals
LOAEL	lowest-observed-adverse-effect level
NIOSH	U.S. National Institute for Occupational Safety and Health
NOAEL	no-observed-adverse-effect level
NCI	National Cancer Institute
NRC	National Research Council
NTP	National Toxicology Program
MMAD	mass median aerodynamic diameter
OSHA	U.S. Occupational Safety and Health Administration
PEL	permissible exposure limit
PNOC	particles not otherwise classified
REGL	repeated exposure guidance level (referred to as permissible exposure guidance level in Volume 1)
REL	recommended exposure level
RP	red phosphorus
RP-BR	red phosphorus–butyl rubber
RPEGL	repeated public exposure guidance level (referred to as permissible public exposure guidance level in Volume 1)
SPEGL	short-term public emergency guidance level
STEL	short-term exposure limit
TiO_2	titanium dioxide
TLV	Threshold Limit Value
TWA	time-weighted average
WP	white phosphorus

TOXICITY OF MILITARY SMOKES AND OBSCURANTS

VOLUME 2

Summary

A VARIETY OF smokes and obscurants have been developed and are used to screen armed forces from view, signal friendly forces, and mark positions. Obscurants are anthropogenic or naturally occurring particles suspended in the air that block or weaken transmission of particular parts of the electromagnetic spectrum, such as visible and infrared radiation or microwaves. Fog, mist, and dust are examples of natural obscurants. Smokes are produced by burning or vaporizing some product. Red phosphorus smoke and graphite smoke are examples of anthropogenic obscurants.

The U.S. Army seeks to ensure that exposure to smokes and obscurants during training does not have adverse health effects on military personnel or civilians. To protect the health of exposed individuals, the Office of the Army Surgeon General requested that the National Research Council (NRC) review data on the toxicity of smokes and obscurants and recommend exposure guidance levels for military personnel in training and for the general public residing or working near military-training facilities.

The NRC assigned this project to the Committee on Toxicology (COT), which convened the Subcommittee on Military Smokes and Obscurants. The subcommittee conducted a detailed evaluation of the toxicity of four obscuring smokes: white phosphorus, brass, titanium dioxide, and graphite. The results of the subcommittee's study are presented in this report, which is the second volume in the series. Toxicity data and exposure guidance levels for diesel-fuel, fog-oil, red phosphorus, and hexachloroethane smokes were presented in Volume 1. Seven colored smokes will be reviewed in a subsequent volume.

The Army requested recommendations for four types of exposure guidance levels: (1) emergency exposure guidance levels (EEGLs) for a rare, emergency situation resulting in exposure of military personnel for less than 24 hr; (2) repeated exposure guidance levels (REGLs) for repeated exposure of military personnel during training exercises (referred to as permissible exposure guidance levels in Volume 1); (3) short-term public emergency guidance levels (SPEGLs) for a rare, emergency situation potentially resulting in an exposure of the public to military-training smoke; and (4) repeated public exposure guidance levels (RPEGLs) for repeated exposures of the public residing or working near military-training facilities (referred to as permissible public exposure guidance levels in Volume 1).

RECOMMENDED EXPOSURE GUIDANCE LEVELS FOR MILITARY PERSONNEL

Using the NRC guidelines published in 1986 and 1992 for developing exposure guidance levels, the subcommittee recommends EEGLs and REGLs for the four obscuring smokes. They are summarized in Table S-1 and described in more detail below.

WHITE PHOSPHORUS SMOKE

White phosphorus (WP) smoke is used by the military in mortar and artillery shells and grenades to block the transmission of visible light, infrared light, or microwaves. It is the most effective obscuring smoke to defeat thermal imagery systems. Phosphorus smoke is generated from a phosphorus-containing flammable matrix that burns to form solid particles of phosphorus pentoxide (P_2O_5) in air. P_2O_5 reacts with moisture to form orthophosphoric acid (H_3PO_4). In this report, toxicity data and exposure guidance levels for WP smoke are reported as H_3PO_4 equivalents.

The most sensitive toxic response to acute exposure (one exposure or multiple exposures occurring within a short time, usually 24 hr or less) to WP smoke is respiratory irritation and distress. Such an effect became evident in goats and rats following a 1-hr exposure to H_3PO_4 at 745 and 525 milligrams per cubic meter (mg/m^3), respectively. Human volunteers exposed for 3.5 min to H_3PO_4 at 818 mg/m^3 reported respiratory irritation, tightness of the chest, cough, and difficulty in breathing. The subcommittee considered both the human and the animal data in estab-

TABLE S-1 Recommended EEGLs and REGLs for Four Smokes

Smoke	Exposure Guideline	Exposure Duration	Exposure Guidance Level (mg/m³)
White phosphorus (as orthophosphoric acid)	EEGL	15 min	19
		1 hr	5
		6 hr	0.8
	REGL	8 hr/d, 5 d/wk	0.09
Brass	EEGL	15 min	1.6
		1 hr	0.4
		6 hr	0.07
	REGL	8 hr/d, 5 d/wk	0.001
Titanium dioxide	EEGL	15 min	1800
		1 hr	450
		6 hr	75
	REGL	8 hr/d, 5 d/wk	2
Graphite	EEGL	15 min	880
		1 hr	220
		6 hr	40
	REGL	8 hr/d, 5 d/wk	1

Abbreviations: EEGL, emergency exposure guidance level; REGL, repeated exposure guidance level (referred to as permissible exposure guidance level in Volume 1).

lishing the EEGLs. Using the lowest-observed-adverse-effect level (LOAEL) of 525 mg/m³ identified in rats, the subcommittee applied an uncertainty factor of 10 to extrapolate from the LOAEL to a no-observed-adverse-effect level (NOAEL) and an additional factor of 10 to extrapolate from effects in animals to humans. Assuming that Haber's rule (that is, the product of exposure concentration and time is a constant, $C \times T = k$) applies, the EEGLs for H_3PO_4 were calculated to be 21, 5, and 1 mg/m³ for 15 min, 1 hr, and 6 hr, respectively. The EEGLs derived from human data on H_3PO_4 were calculated to be 19, 5, and 0.8 mg/m³ for 15 min, 1 hr, and 6 hr, respectively. Because human data are available and are supported by the animal data, the subcommittee recommends using the EEGLs derived from human data.

To recommend a REGL for H_3PO_4, the subcommittee used the NOAEL of 280 mg/m³ identified in a study in which rats were exposed to H_3PO_4

for 15 min per day, 5 days per week for 13 weeks. An uncertainty factor of 10 was used to extrapolate from animal to human data. Assuming Haber's rule applies, derivation of an 8-hr exposure value from a 15-min exposure yields a value of 0.9 mg/m^3. Applying an uncertainty factor of 10 to extrapolate from subchronic (multiple exposures occurring usually over 3 months) to chronic (multiple exposures occurring over a significant fraction of the animal's lifetime) exposure yields a value of 0.09 mg/m^3. Therefore, the subcommittee recommends a REGL for WP smoke (expressed as H_3PO_4) of 0.09 mg/m^3 for 8 hr per day, 5 days per week.

BRASS SMOKE

Explosion of grenades containing brass flakes composed of 70% copper and 30% zinc results in the release of these flakes into the atmosphere to block detection of infrared waves for thermal imagery systems.

Inhalation exposures to high concentrations of brass flakes are acutely lethal to guinea pigs, rats, and mice. Pulmonary inflammation can be produced in rats exposed at a concentration of 10 mg/m^3 for 4 hr. The response is reversible, and clearance of the flakes from the lungs appears to be rapid.

Using the NOAEL of 1 mg/m^3 identified in rats, and applying Haber's rule and an uncertainty factor of 10 to extrapolate from animals to humans, the subcommittee estimates the 15-min, 1-hr, and 6-hr EEGLs for brass flakes to be 1.6 mg/m^3, 0.4 mg/m^3, and 0.07 mg/m^3, respectively.

Multiple exposures to brass flakes are more toxic than single exposures. Humans who are occupationally exposed to substantial amounts of brass dust exhibit chronic bronchitis and other severe respiratory problems, although most such workers are also likely to be exposed to additional respiratory irritants. A concentration in humans known not to be associated with respiratory disease is about 0.001 mg/m^3. The subcommittee recommends that value as the REGL for brass smoke. Data on subchronic exposure of rats to brass flakes also support that value.

TITANIUM DIOXIDE SMOKE

Titanium dioxide (TiO_2) is the proposed major component of a training grenade—XM82— under development by the U.S. Army Chemical Research Development and Engineering Center. Explosion of grenades containing TiO_2 will release TiO_2 particles that block detection of light waves in the visible portion of the electromagnetic spectrum.

No acute human toxicity data are available on TiO_2. On the basis of animal studies, the potential for lethality from a single exposure is minimal. A single 30-min exposure at 1,240 mg/m^3 was not toxic to rats. Therefore, acute inhalation of TiO_2 is considered to have minimal toxicity.

The subcommittee's recommended EEGLs for TiO_2 are based on estimates of the acute exposure concentration of TiO_2 that are below the lung concentration associated with impaired particle clearance and adverse lung effects. On the basis of long-term inhalation studies in rodents, a lung concentration of TiO_2 at less than 4 milligrams per gram (mg/g) of lung would not be expected to alter alveolar particle clearance. To account for potentially increased effects due to the high concentration used in acute inhalation exposures, the subcommittee reduced the concentrations of 4 mg/g of lung by a factor of 10 and estimated the maximal 15-min, 1-hr, and 6-hr exposure concentrations that would result in a lung concentration of less than 0.4 mg/g of lung. The subcommittee calculated that a 15-min exposure at 1,800 mg/m^3 would result in a lung concentration of less than 0.4 mg/g of lung. Therefore, the subcommittee recommends that the 15-min EEGL for TiO_2 smoke be 1,800 mg/m^3. Extrapolating from the 15-min EEGL, the subcommittee recommends a 1-hr EEGL of 450 mg/m^3 and a 6-hr EEGL of 75 mg/m^3 for TiO_2 smoke.

To develop a REGL for TiO_2, the subcommittee estimated a chronic human exposure concentration that would not overload alveolar particle-clearance mechanisms and lead to adverse effects resulting from accumulation of particles in the lung. As noted above, a lung concentration of less than 4 mg/g of lung would not alter alveolar particle clearance. The subcommittee calculated that exposure to TiO_2 at 2.0 mg/m^3 for 8 hr per day, 5 days per week is the maximal exposure concentration that would result in a lung concentration of less than 4 mg/g of lung. Therefore, the subcommittee recommends a REGL for TiO_2 smoke of 2.0 mg/m^3 for 8 hr per day, 5 days per week.

GRAPHITE SMOKE

Graphite flakes are used by the military to block electromagnetic waves that an enemy might detect and use to target troops in the field.

Available acute toxicity data suggest that graphite does not cause adverse health effects in humans. Humans exposed chronically to graphite dust might develop graphite pneumoconiosis. However, this effect is believed to be due to impurities, particularly silica, in the graphite that is mined. Acute and subchronic inhalation studies of pure graphite in animals showed only minimal inflammatory reactions in the respiratory tract. The inflammation was fully reversible most of the time. No deaths occurred even at the highest concentrations tested. If graphite particles persist in the lung, however, epithelia of the terminal bronchioles and alveoli show signs of hyperplasia, and graphite-containing granulomas develop in lymphoid tissue.

The subcommittee used an approach similar to that used for TiO_2 to develop EEGLs for graphite. Data from chronic inhalation studies in rats show that a lung concentration of less than 2 mg/g of lung should not alter alveolar particle clearance or produce adverse effects in the lung. To account for potentially increased effects due to the high concentrations used in acute inhalation exposures, the subcommittee reduced the concentration of 2 mg/g of lung by a factor of 10 and estimated the maximal 15-min, 1-hr and 6-hr exposure concentrations that would result in a lung concentration of less than 0.2 mg/g of lung for a single exposure. The subcommittee calculated that a 15-min exposure to graphite at approximately 880 mg/m^3 would be the maximal concentration resulting in a lung concentration of less than 0.2 mg/g of lung. Extrapolating from the 15-min EEGL, the subcommittee recommends a 1-hr EEGL of 220 mg/m^3 and a 6-hr EEGL of 40 mg/m^3 for graphite smoke.

To calculate a REGL for graphite, the subcommittee recommends that it be based on the maximal chronic exposure concentration that would not overload lung particle-clearance mechanisms, resulting in accumulations of particles. As discussed above, a lung concentration of less than 2 mg/g of lung would not alter alveolar particle clearance. The subcommittee calculated that chronic exposure to graphite at 1.0 mg/m^3 would result in a lung concentration of less than 2 mg/g of lung. Thus, the subcommittee recommends a REGL for graphite smoke of 1.0 mg/m^3 for 8 hr per day, 5 days per week.

RECOMMENDED EXPOSURE GUIDANCE LEVELS FOR COMMUNITIES NEAR MILITARY-TRAINING FACILITIES

SPEGLs and RPEGLs are established to ensure the protection of communities near military facilities. In recommending SPEGLs and RPEGLs, the subcommittee assumed that the general population includes susceptible subpopulations, such as the elderly, the chronically ill, pregnant women, infants, children, developing fetuses, and embryos. In the absence of direct information on the toxicity of the smokes in susceptible subpopulations, an uncertainty factor of 10 was used to extrapolate EEGLs and REGLs derived for a population of healthy adults in the military to concentrations protective of more-susceptible subpopulations. The SPEGLs and RPEGLs recommended by the subcommittee are shown in Table S-2 and are 0.1 times the corresponding EEGLs and REGLs shown in Table S-1.

TABLE S-2 Recommended SPEGLs and RPEGLs for Four Smokes

Smoke	Exposure Guideline	Exposure Duration	Exposure Guidance Level (mg/m^3)
White phosphorus (as orthophosphoric acid)	SPEGL	15 min	1.9
		1 hr	0.5
		6 hr	0.08
	RPEGL	8 hr/d, 5 d/wk	0.009
Brass	SPEGL	15 min	0.16
		1 hr	0.04
		6 hr	0.007
	RPEGL	8 hr/d, 5 d/wk	0.0001
Titanium dioxide	SPEGL	15 min	180
		1 hr	45
		6 hr	7.5
	RPEGL	8 hr/d, 5 d/wk	0.2
Graphite	SPEGL	15 min	88
		1 hr	22
		6 hr	4.0
	RPEGL	8 hr/d, 5 d/wk	0.1

Abbreviations: SPEGL, short-term public emergency guidance level; RPEGL, repeated public exposure guidance level (referred to as permissible public exposure guidance level in Volume 1).

1

Introduction

EVER SINCE smokeless powder replaced black powder as the standard propellant for guns and firearms, the armed forces have sought methods to blanket battlefields by creating a haze similar to that created by black powder. A variety of smokes and obscurants has been developed and used in wartime operations for screening armed forces from view, deceiving the enemy, signaling friendly forces, and marking positions. White and gray smokes are deployed in grenades to cover or screen individual vehicles, and colored smokes are used to mark specific locations. Obscurants are anthropogenic or naturally occurring particles suspended in air that block or weaken the transmission of particular parts of the electromagnetic spectrum, such as visible and infrared radiation or microwaves. Smokes and obscurants are used by the armed forces to achieve tactical goals during wartime. To ensure defense preparedness, large quantities of smokes and obscurants also are used in military training.

THE SUBCOMMITTEE'S TASK

To ensure that exposure to smokes and obscurants during combat training will not have adverse health effects on military personnel, the Office of the Army Surgeon General requested the National Research Council (NRC) to review the data on the toxicity of military smokes and obscurants and recommend exposure guidance levels for military personnel during combat training and for the general public residing or working near military-training facilities.

The NRC assigned this project to the Committee on Toxicology (COT), which convened the Subcommittee on Military Smokes and Obscurants. For this report, the subcommittee evaluated four obscuring smokes: white phosphorus, brass, graphite, and titanium dioxide. Exposure guidance levels for four other obscuring smokes—fog oil, diesel fuel, red phosphorus, and hexachloroethane—were presented in Volume 1 of *Toxicity of Military Smokes and Obscurants* (1997), and seven signaling smokes will be presented in a subsequent volume.

The task of the subcommittee was to review the health effects associated with exposure to the smokes and obscurants and to recommend four exposure guidance levels: (1) emergency exposure guidance levels (EEGLs) for a rare, emergency situation resulting in an exposure of military personnel for less than 24 hr; (2) repeated exposure guidance levels (REGLs) for repeated exposure of military personnel during training (referred to as permissible exposure guidance level in Volume 1); (3) short-term public emergency guidance levels (SPEGLs) for a rare, emergency situation potentially resulting in an exposure of the public to a military-training smoke; and (4) repeated public exposure guidance levels (RPEGLs) for possible repeated exposures of the public residing or working near military-training facilities (referred to as permissible public exposure guidance level in Volume 1). All four guidance levels should take into account embryo and fetal development and reproductive toxicity in men and women. In addition, exposures of potentially susceptible subpopulations (e.g., ill or elderly persons and children) are considered in the SPEGL and RPEGL.

DEFINITIONS OF EXPOSURE GUIDANCE LEVELS

For each smoke reviewed in this report, the subcommittee recommends the following exposure guidance levels: EEGLs, REGLs, SPEGLs, and RPEGLs. Table 1-1 summarizes the definitions of these exposure guidance levels and a more detailed description for each is given below.

An EEGL is defined as a concentration of a substance in air (as a gas, vapor, or aerosol) that will permit continued performance of specific tasks during emergency exposures lasting up to 24 hr—an occurrence expected to be infrequent in the lifetime of a person (NRC 1986, 1992a). "Emergency" connotes a rare and unexpected situation with potential for significant loss of life, property, or mission accomplishment if not controlled. An EEGL, a single ceiling-exposure concentration for a specified

> **TABLE 1-1 DEFINITIONS OF EXPOSURE GUIDANCE LEVELS FOR MILITARY SMOKES AND OBSCURANTS**
>
> EEGL   Emergency exposure guidance level for a rare, emergency situation resulting in an unanticipated exposure of military personnel for less than 24 hr.
>
> REGL   Repeated exposure guidance level for repeated exposure of military personnel during training. This exposure guidance level was referred to as permissible exposure guidance level in Volume 1.
>
> SPEGL Short-term public emergency guidance level for a rare, emergency situation potentially resulting in an exposure of the general public to military-training smoke.
>
> RPEGL Repeated public exposure guidance level for possible repeated exposures of the general public residing or working near military-training facilities. This exposure guidance level was referred to as permissible public exposure guidance level in Volume 1.

duration, specifies and reflects the subcommittee's interpretation of available information in the context of an emergency.

An EEGL is acceptable only in an emergency, when some risks or some discomfort must be endured to prevent greater risks (such as fire, explosion, or massive release). Exposure at the EEGL might produce such effects as increased respiratory rate, headache, mild central-nervous-system effects, and respiratory-tract or eye irritation. The EEGL should prevent irreversible harm. Even though some reduction in performance is permissible, it should not prevent proper responses to the emergency (such as shutting off a valve, closing a hatch, or using a fire extinguisher). For example, in normal work situations, a degree of upper-respiratory-tract irritation or eye irritation causing discomfort would not be considered acceptable; during an emergency, it would be acceptable if it did not cause irreversible harm or seriously affect judgment or performance. The EEGL for a substance represents the subcommittee's judgment based on evaluation of experimental and epidemiological data, mechanisms of injury, and, when possible, operating conditions in which an emergency exposure might occur, as well as consideration of U.S. Department of Defense (DOD) goals and objectives. EEGLs were developed for military use and are intended for healthy military personnel. Therefore, they are not directly applicable to general populations consisting of elderly, very young, and ill persons.

A SPEGL is defined as a concentration of a substance in air that is acceptable for an unpredicted, single or rare, short-term emergency exposure of the general public. The SPEGL takes into account the likely wide range of susceptibility among individuals in the general public, including potentially susceptible subpopulations, such as children, the elderly, and persons with serious debilitating diseases. Effects of exposure on the developing embryo and fetus and on the reproductive capacity of men and women also are considered in setting a SPEGL.

For purposes of assessing military smokes and other obscurants for the Army, the subcommittee developed two additional guidance levels, REGLs and RPEGLs. The subcommittee defines a REGL as the concentration of a substance in air to which healthy military personnel can be exposed repeatedly, up to a specified total exposure on a weekly basis (usually 8 hr per day, 5 days per week), for several years without experiencing adverse health effects or degradation in performance.

The subcommittee defines a RPEGL as the concentration of a substance in air to which the general public can be exposed repeatedly without experiencing any adverse health effects or discomfort. RPEGLs, like SPEGLs, take into account the likely wide range of susceptibility among individuals in the general public, including potentially susceptible subpopulations (children, the elderly, and the chronically ill), the developing embryo and fetus, and reproductive capacity in men and women.

Exposure guidance levels developed by the subcommittee can be compared with other potentially useful exposure limits, such as the Threshold Limit Value (TLV) time-weighted averages (TWAs) and short-term exposure limits (STELs) recommended by the American Conference of Governmental Industrial Hygienists (ACGIH) for permissible workplace exposures (ACGIH 1991, 1995). These guidelines are developed for daily occupational exposures of healthy workers. The Occupational Safety and Health Administration (OSHA) is responsible for promulgating and enforcing health standards in the majority of the work environments. These legally binding standards are referred to as permissible exposure limits (PELs; U.S. Department of Labor 1997), and most follow the National Institute of Occupational Safety and Health (NIOSH) recommended exposure levels (RELs; EPA 1987). The OSHA and ACGIH values are not relevant for the general public.

APPROACH TO DEVELOPING EXPOSURE GUIDANCE LEVELS

The NRC has published guidelines for developing EEGLs, SPEGLs, and other exposure guidance levels for continuous or repeated exposures (NRC 1986, 1992a,b, 1996). For purposes of assessing military smokes and other obscurants, the subcommittee developed comparable procedures for developing REGLs and RPEGLs. The steps in developing exposure guidance levels are similar for EEGLs, SPEGLs, REGLs, and RPEGLs; the differences reflect attributes of the exposed populations and the duration and frequency of exposure. The remainder of this section reviews the steps for developing an EEGL (NRC 1986) and then explains how procedures differ for the remaining three types of exposure guidance levels.

EMERGENCY EXPOSURE GUIDANCE LEVELS

The first step in developing an EEGL is to review all available toxicology information and any documentation for exposure limits proposed by ACGIH and regulatory agencies. Acute toxicity is the primary basis for establishing an EEGL. All end points are evaluated, and the most important are selected. In general, EEGLs reflect experimental and clinical observations and are based on epidemiological, physiological, and toxicological data on animals and humans; both immediate and delayed health effects are considered. Special attention is given to training and battlefield conditions that are of concern to the military. If there is any evidence that the substance under consideration is carcinogenic in either animals or humans, a cancer risk assessment is performed to estimate the possible potency of the substance as a carcinogen. The approach used to estimate potency is developed case by case, depending on available data and plausible mechanisms of action.

In evaluating the noncancer health effects, the subcommittee first assesses relevant toxicological information to determine the no-observed-adverse-effect level (NOAEL) of each smoke for the most sensitive end point. The NOAEL is the highest concentration at which no adverse toxic effect is indicated by the available data. If a NOAEL cannot be determined from the data, the lowest-observed-adverse-effect level (LOAEL) is determined. The LOAEL is the lowest concentration at which an adverse effect is seen in either human or laboratory animal studies. To estimate a NOAEL from a LOAEL, the subcommittee generally divides the LOAEL by a default uncertainty factor of 10, following the recommendations of the U.S. Environmental Protection Agency (EPA 1994). When NOAEL values obtained from laboratory animals are used to estimate exposure guidelines for humans, the subcommittee adopts the NRC (1977) Safe Drinking Water Committee default assumption that humans are 10-fold more sensitive than animals unless data are available that justify using a different assumption.

SHORT-TERM PUBLIC EMERGENCY GUIDANCE LEVELS

SPEGLs are generally set at 0.1 times the EEGL to protect susceptible subpopulations, including infants, the elderly, the chronically ill, and the developing embryo or fetus (NRC 1986).

REPEATED EXPOSURE GUIDANCE LEVELS AND REPEATED PUBLIC EXPOSURE GUIDANCE LEVELS

Although the NRC has not published guidelines for developing REGLs and RPEGLs, the subcommittee follows the same approach as that recommended for EEGLs and SPEGLs; modifications reflect the repeated nature of exposures. In contrast to EEGLs and SPEGLs, the subcommittee uses chronic toxicity as the primary basis for establishing REGLs and RPEGLs. Haber's rule might not be applicable to the longer-term exposure durations. If there is any evidence that the substance is carcinogenic in either animals or humans, the subcommittee estimates the possible potency of the compound as a carcinogen. DOD can use the potency value when comparing risks associated with different concentrations of exposure with risks incurred by personnel wearing masks or not using the obscurant. The subcommittee generally sets RPEGLs at 0.1 times the REGLs to protect more susceptible subpopulations in the general public.

CONFIDENCE LEVEL IN USING THE PRODUCT OF UNCERTAINTY FACTORS

In recommending exposure guidance levels for the various obscurants, the subcommittee has had to rely on the use of uncertainty factors in an attempt to account for various deficiencies in the toxicology data base for each obscurant. Those deficiencies include absence of a NOAEL, inadequate toxicity information on humans, and absence of information on potential susceptible subpopulations. Default 10-fold uncertainty factors are used in each case (i.e., LOAEL to NOAEL, extrapolation of animal data to humans, and variability in human susceptibility) but might be reduced if, for example, adjustments for species differences in lung dosimetry are made, as is done in the reference concentration method (EPA 1994). In that case, a lower (e.g., 3-fold) uncertainty factor might be applied. Potential sensitivity of the developing embryo and fetus and the reproductive system are accounted for in part by an intraspecies-human-variability uncertainty factor. Deficiencies in reproductive and developmental toxicity data are sometimes accounted for by applying a data-base uncertainty factor in addition to the intraspecies uncertainty factor, because reproductive and developmental end points might be more sensitive than other types of end points (Dourson et al. 1992). Because most smokes are primarily pulmonary toxicants, the subcommittee decided that it was unnecessary to apply an additional uncertainty factor to smokes lacking data on reproductive and developmental toxicity.

Thus, the recommended guidance levels developed are not precise, because the exact magnitude of the uncertainties is unknown. The product of the uncertainty factors results in high confidence that the overall factor is large enough to protect susceptible subpopulations adequately from long-term exposures. There is some conjecture on whether multiplying several uncertainty factors might result in an overly conservative guidance level, but that is difficult to resolve in the absence of more specific data. Several investigators have addressed this issue by reviewing data on variability in susceptibility among species and within the human population (Calabrese 1985; Hattis et al. 1987; Dourson et al. 1992; Burmaster and Harris 1993; Calabrese and Gilbert 1993; Bogen 1994; Baird et al. 1996; Dourson et al.1996; Renwick and Lazarus 1998). EPA recognizes this problem and has suggested using a maximum of 10,000 for the total uncertainty factor (Dourson 1994).

ORGANIZATION OF THE REPORT

This report is organized into four chapters, one for each of the obscuring smokes evaluated in this volume. For each smoke, the chapter presents background information on military applications and physical and chemical properties of the smoke. Each chapter also includes a discussion of toxicokinetics and a summary of the available toxicity data on the smoke. Following a description of existing recommended exposure limits, each chapter presents the subcommittee's evaluation of the toxicity data and the development of the exposure guidance levels. Sections on research needs and references conclude each chapter.

REFERENCES

ACGIH (American Conference of Governmental Industrial Hygienists). 1991. Documentation of the Threshold Limit Values and Biological Exposure Indices, 6th Ed. American Conference of Governmental Industrial Hygienists, Cincinnati, Ohio.

ACGIH (American Conference of Governmental Industrial Hygienists). 1995. 1995-1996 Threshold Limit Values and Biological Exposure Indices. American Conference of Governmental Industrial Hygienists, Cincinnati, Ohio.

Baird, S.J.S., J.T. Cohen, J.D. Graham, A.I. Shlyakhter, and J.S. Evans. 1996. Noncancer risk assessment: A probabilistic alternative to current practice. Hum. Ecol. Risk Assess. 2:79-102.

Bogen, K.T. 1994. A note on compounded conservatism. Risk Anal. 14:379-381.

Burmaster, D.E., and R.H. Harris. 1993. The magnitude of compounding conservatisms in superfund risk assessments. Risk Anal. 13:131-134.

Calabrese, E.J. 1985. Uncertainty factors and interindividual variation. Regul. Toxicol. Pharmacol. 5:190-196.

Calabrese, E.J., and C.E. Gilbert. 1993. Lack of total independence of uncertainty factors (UFs): Implications for the size of the total uncertainty factor. Regul. Toxicol. Pharmacol. 17:44-51.

Dourson, M.L., L.A. Knauf, and J.C. Swartout. 1992. On Reference Dose (RfD) and its underlying toxicity data base. Toxicol. Ind. Health 8:171-189.

Dourson, M.L. 1994. Methods for establishing oral reference doses (RfDs). Pp. 51-61 in Risk Assessment of Essential Elements, W. Mertz, C.O. Abernathy, and S.S. Olin, eds. Washington, D.C.: ILSI Press.

Dourson, M.L., S.P. Felter, and D. Robinson. 1996. Evolution of science-based uncertainty factors in noncancer risk assessment. Regul. Toxicol. Pharmacol. 24:108-120.

Eaton, J.C., and J.Y. Young. 1989. P. 11 in Medical Criteria for Respiratory Protection in Smoke: The Effectiveness of the Military Protective Mask. Tech. Rep. 8902. U.S. Army Biomedical Research and Development Laboratory, Fort Detrick, Frederick, Md.

Eckelbarger, M.G. 1985. Smoke Safety. Message from the Director of Army Safety, U.S. Department of the Army, Washington, D.C.

EPA (U.S. Environmental Protection Agency). 1987. Technical Guidance for Hazards Analysis: Emergency Planning for Extremely Hazardous Substances. Prepared by the U.S. Environmental Protection Agency in conjunction with the Federal Emergency Management Agency and the U.S. Department of Transportation, Washington, D.C. Available from NTIS, Springfield, Va., Doc. No. PB93-206910.

EPA (U.S. Environmental Protection Agency). 1994. Methods for Derivation of Inhalation Reference Concentrations and Application of Inhalation Dosimetry. EPA 600/8-90/066F. U.S. Environmental Protection Agency, Environmental Criteria and Assessment Office, Research Triangle Park, N.C.

Hattis, D., L. Erdreich, and M. Ballew. 1987. Human variability in susceptibility to toxic chemicals: A preliminary analysis of pharmacokinetic data from normal volunteers. Risk Anal. 7:415-426.

NRC (National Research Council). 1977. Drinking Water and Health. Vol. 1. Washington, D.C.: National Academy Press.

NRC (National Research Council). 1986. Criteria and Methods for Preparing Emergency Exposure Guidance Level (EEGL), Short-term Public Emergency Guidance Level (SPEGL), and Continuous Exposure Guidance Level (CEGL) Documents. Washington, D.C.: National Academy Press.

NRC (National Research Council). 1992a. Guidelines for Developing Community Emergency Exposure Levels for Hazardous Substances. Washington, D.C.: National Academy Press.

NRC (National Research Council). 1992b. Guidelines for Developing Spacecraft Maximum Allowable Concentrations for Space Station Contaminants. Washington, D.C.: National Academy Press.

NRC (National Research Council). 1996. Toxicity of Alternatives to Chlorofluorocarbons: HFC-134a and HCFC-123. Washington, D.C.: National Academy Press.

NRC (National Research Council). 1997. Toxicity of Military Smokes and Obscurants. Washington, D.C.: National Academy Press.

Renwick, A.G., and N.R. Lazarus. 1998. Human variability and noncancer risk assessment—An analysis of the default uncertainty factor. Regul. Toxicol. Pharmacol. 27:3-20.

U.S. Department of Labor. 1997. Occupational Safety and Health Standards. Air Contaminants. Code of Federal Regulations, Title 29, Part 1910, Section 1910.1000. Washington, D.C.: U.S. Government Printing Office.

2

White Phosphorus Smoke

BACKGROUND INFORMATION

OBSCURANTS, such as phosphorus smokes, are effective in blocking the transmission of a particular part of the electromagnetic spectrum, such as visible light, infrared light, or microwaves. Military application of phosphorus smokes for screening during a military operation can make use of either white phosphorus (WP) or red phosphorus (RP). WP is the most effective smoke agent to defeat thermal imagery systems. It can be absorbed by the inhalation, ingestion, or dermal routes; however, the primary route is inhalation. This chapter provides an evaluation of the health effects of elemental WP and WP smoke. Volume 1 of this series reviewed the toxicity associated with the inhalation of red phosphorus smoke (NRC 1997).

MILITARY APPLICATIONS

The military uses smokes, such as WP, to protect friendly forces, support deception operations, identify enemy targets and tactical locations, and obscure certain reconnaissance activities, surveillance, and targets from enemy forces. WP is used in mortar and artillery shells and in grenades. WP used as a military obscurant is impregnated in a felt matrix and is referred to as white phosphorus–felt. Unprotected troops in training exercise or combat are likely to inhale the smoke from a detonated grenade. The U.S. Environmental Protection Agency (EPA) estimated that an exposure concentration of WP could reach 146 milligrams per cubic meter (mg/m^3) as phosphorus pentoxide (P_2O_5) (202 mg/m^3 as orthophosphoric acid (H_3PO_4)) 100 m downwind from deployment and about

1.0 mg/m³ as P₂O₅ (1.4 mg/m³ as H₃PO₄) 5,000 m downwind (EPA 1990). EPA does not expect community exposures to be severe at a distance of greater than 300 m; however, particularly susceptible individuals might experience respiratory irritation even at a distance of 5,000 m (EPA 1990).

PHYSICAL AND CHEMICAL PROPERTIES

CAS no.: 12185-10-3
Synonyms: yellow phosphorus, phosphorus tetramer
Chemical formula: P_4
Chemical weight: 123.9
Physical state: waxy solid
Density at 20°C: 1.82 g/cm³
Melting point: 44.1°C
Boiling point: 280°C
Vapor pressure (20°C): 0.025 mm Hg
Flash point: spontaneous in air
Solubility in water: 3 mg/L
Conversion factor: 1 ppm = 5.150 mg/m³ at 20°C

Elemental phosphorus exists in a number of allotropic forms. WP is an inorganic chemical that has a slight yellow color caused by traces of red phosphorus impurities and is often referred to as yellow phosphorus. WP is poorly soluble in water but is soluble in nonpolar organic solvents, such as benzene. WP can react with water to form a gaseous compound, phosphine (PH_3), which is toxic to the central nervous system and the liver. PH_3 will rapidly volatilize from water into air because of its low water solubility and high vapor pressure. In the air, PH_3 is converted to less harmful chemicals.

OCCURRENCE AND USE

WP is a form of phosphorus that does not occur naturally. It is manufactured from naturally occurring phosphate rocks (ATSDR 1997). WP has been used in producing phosphoric acid, fertilizers, additives in food and beverages, cleaning compounds, fireworks, and smoke bombs. WP can enter the environment in areas of elemental phosphorus production sites and hazardous dump sites, near industries that use it, from accidental spills, and during military use in training and warfare. Because of the high reactivity of WP, it usually is not found far from the source of contamination.

COMBUSTION PRODUCTS

When munitions containing WP are fired, they burn and produce smoke. The combustion of WP will produce smoke made up of various oxides of phosphorus, including P_2O_5 and phosphorus trioxide (P_4O_6). These oxides react rapidly with moisture to form a number of transformation products, such as H_3PO_4 and pyrophosphoric acid ($H_4P_2O_7$) (Table 2-1; Brazell et al. 1984) and about 10% unburned phosphorus (Spanggord et al. 1985; ATSDR 1997). Organic compounds (concentrations in parts per billion) and some inorganic gases might be present, but only at trace levels. Because WP is not likely to persist long in air, a majority of phosphorus compounds released and dispersed in air during military use of smokes are likely to be deposited as phosphoric acid or phosphates on land and water (EPA 1990). The smoke particle diameter is about 1 micrometer (μm) by count, 98% of the particles are below 2μm in diameter (Katz et al. 1981).

The chemical characteristics of WP and red-phosphorus–butyl-rubber (RP–BR) smokes are similar; both are primarily phosphoric acids, present as a complex mixture of polymeric forms. WP reacts more rapidly in air ($t_{1/2}$ = 5 min), and RP–BR is more persistent in air ($t_{1/2}$ = 1.8 years) (Spanggord et al. 1985; EPA 1990). However, if the particulate phosphorus is coated with a protective layer of oxide, further oxidation might not occur, increasing the lifetime of the elemental phosphorus in the air and on the ground after deposition (ATSDR 1997).

MEASUREMENT

Different methods have been used to measure or estimate the concentration of WP smoke in the air. In the toxicological literature, some authors have reported concentrations in terms of P_2O_5 (molecular weight (mol wt) 141.94; which exists as P_4O_{10}; mol wt 283.89) equivalents (White

TABLE 2-1 Composition of Phosphoric Acids in White Phosphorus Smokes Produced from Static Burn

Component	Chemical Formula	Composition by weight (%)
Orthophosphoric acid	H_3PO_4	23.8
Pyrophosphate	$H_4P_2O_7$	26.6
Tripolyphosphate	$H_5P_3O_{10}$	16.3
Tetrapolyphosphate	$H_6P_4O_{13}$	11.3
P_5-P_{13}	$H_{n+2}P_nO_{3n+1}$	22.0
Higher polyphosphates	$H_{n+2}P_nO_{3n+1}$	Low

Source: Brazell et al. 1984.

and Armstrong 1935) and others in terms of H_3PO_4 (mol wt 98.00) equivalents (Brown et al. 1980, 1981; Starke et al. 1982). Concentrations in the air can also be estimated by subtracting the residue remaining from the mass of a quantity burned in an enclosed area of known volume.

To estimate P_2O_5, Armstrong and White (1935) captured the phosphorus smoke on dry asbestos filters and brought the residue into solution by rinsing with water and then boiling the solution for 30 min. They found that boiling for 30 min yields practically complete hydration of P_2O_5 to H_3PO_4, which then can be measured by titration with a phenolphthalein indicator. Brown et al. (1980, 1981), Weimer et al. (1977), and presumably Starke et al. (1982) boiled the solution for 10 min to convert the phosphorus acids to H_3PO_4 and, using a pH meter as an indicator, titrated the solution to a pH of 9.6 with sodium hydroxide. The molecular weight of H_3PO_4 is 98.00, or 32.66 mg/mole of hydrogen atoms. Brown et al. (1981) stated that they multiplied the normality of the acid by 32.66 mg. That calculation assumes that all three hydrogens on the acid are ionized at pH 9.6; that is not the case. The dissociation constants, pK, for H_3PO_4, steps 1, 2, and 3, are 2.12 (at 25°C), 7.21 (at 25°C), and 12.67 (at 18°C), respectively (Lide 1991). Thus, at pH 9.6, only two of the hydrogen ions are dissociated. Hence, the subcommittee multiplied the H_3PO_4 equivalents estimated by Brown et al. (1980, 1981) and Starke et al. (1982) by 3/2 to correct for the undissociated HPO_4^{-2} not accounted for in the original estimates. For example, if Brown et al. (1981) reported that an average exposure concentration of H_3PO_4 equivalents was 589 mg/m^3, that concentration was multiplied by 3/2 to yield H_3PO_4 equivalents at 884 mg/m^3.

For the sake of consistency in this report, toxicity data for WP smoke are reported as H_3PO_4 equivalents when possible. Exposure to H_3PO_4 would occur because it is a combustion endproduct. To convert P_2O_5 equivalents to H_3PO_4 equivalents, the concentration of the former is multiplied by a factor of 195.99/141.94 or 1.381. Each P_2O_5 molecule (mol wt 141.94) produces two H_3PO_4 molecules (mol wt 98.00) when hydrated. Thus, 141.94 grams (g) of P_2O_5 is equivalent to 195.99 g of H_3PO_4.

TOXICOKINETICS

The toxicokinetics and ultimate fate of inhaled WP in the body are unknown. No studies have been conducted on the absorption, distribution, metabolism or excretion of WP following inhalation of WP smoke by either humans or animals. Although it is not known whether inhaled WP enters the blood, the oxides and acids of WP that occur in the smoke might be absorbed, but to an unknown extent, and possibly distributed systemically.

TOXICITY SUMMARY:
WHITE PHOSPHORUS AND WHITE PHOSPHORUS SMOKE

WHITE PHOSPHORUS: HUMAN AND ANIMAL STUDIES

Generally, WP is highly toxic when individuals accidentally or intentionally ingest a single dose, in contrast to RP, which is insoluble and not absorbed when ingested (Simon and Pickering 1976; Wasti et al. 1978). Oral ingestion of WP in humans can be lethal at concentrations of 1 mg/kilogram (kg) of body weight (H_3PO_4 at 3.2 mg/kg). Amounts as low as 0.2 mg/kg (H_3PO_4 at 0.63 mg/kg) can cause severe effects (Watsi et al. 1978; Yon et al. 1983). Fatal or near-fatal human exposures have occurred as a consequence of oral ingestion during a suicide attempt or as a consequence of dermal burns during munitions explosions. Systemic effects following oral ingestion in humans and animals usually begin with severe gastrointestinal distress resulting from irritation of the gastrointestinal lining. That can be followed (during the next 3 weeks) by potentially life-threatening organ impairments such as cardiac arrest, fatty infiltration of the liver and kidneys, and hepatomegaly. Hematological and neurological effects have also been observed.

No studies have investigated the lethal effects of inhalation of WP by humans nor have they reported possible effects from short-term exposures; such effects might include respiratory, cardiovascular, gastrointestinal, hematological, hepatic, renal, dermal and ocular, immunological, neurological, reproductive, developmental, genotoxic, and carcinogenic effects.

However, long-term occupational exposure to airborne phosphorus vapors present in the atmosphere of a factory has been found to produce a degenerative condition resulting in necrosis of the jaw, known as phossy jaw, in some workers. The effects can be extreme, involving severe necrosis of the entire oral cavity, including the soft tissue, teeth, and bones (Heimann 1946 and Hughes et al. 1962, as cited in ATSDR 1997). Massive life-threatening infections can follow. The effect is thought to be a result of direct contact of the inhaled white phosphorus particles with the tissues in the mouth.

No laboratory animal studies have investigated the gastrointestinal, musculoskeletal, dermal and ocular, immunological, developmental, reproductive, neurological, genotoxic, or carcinogenic effects of inhaled WP. One study found the lowest published lethal concentration of WP for mice to be 500 mg/m^3 (H_3PO_4 at 1,600 mg/m^3) for 10 min (Lee et al. 1975). In another study, rabbits exposed for 30 min to WP at concentrations of 150-160 mg/m^3 (H_3PO_4 at 470-500 mg/m^3) for 60 days exhibited a decrease in hemoglobin and erythrocyte counts (Maruo 1955). A study that placed rats for an intermediate duration in a phosphorus factory that was reported to contain WP and certain inorganic compounds described degeneration of the tongue and oral mucous of the cheek, gum, and hard palate (Ruzuddinov and Rys Uly 1986, as cited in ATSDR 1997). The exposure duration and concentrations were not reported.

WP has been tested for mutagenicity in the Ames test. WP in water at a concentration of 100 microliters (μL) per plate produced no mutagenic activity in *Salmonella* strains TA100, TA1535, TA98, TA1537, and TA1538 in either the presence or the absence of metabolic activation (Ellis et al. 1978, as cited in EPA 1990).

WHITE PHOSPHORUS SMOKE: EFFECTS IN HUMANS

Relatively little information has been reported on human responses to inhalation of WP smoke. Exposure of 108 men to WP smoke at 87-1,770 mg/m^3 resulted in coughing and irritation of the throat (Cullumbine 1944, as cited in Wasti et al. 1978). The method used to measure the smoke concentration and the length of exposure were not reported. From those data, Cullumbine (1944, as cited in Wasti et al. 1978) estimated that the minimal exposure concentration causing coughing and throat irritation is about 700 mg/m^3 for working individuals and 1,000 mg/m^3 for individuals at rest.

A number of studies were conducted by White and Armstrong in 1935 with human volunteers. In most of those studies, the individuals were placed in a chamber, and then WP smoke was introduced. Male subjects were exposed to WP smoke with average concentrations of P_2O_5 at 188-514 mg/m^3 for 2 to 15 min (White and Armstrong 1935). At the lowest concentration (P_2O_5 at 188 mg/m^3 or H_3PO_4 at 259 mg/m^3), a 5-min exposure resulted in 50% of the individuals reporting respiratory distress, coughing, congestion, and throat irritation. At the highest concentration (P_2O_5 at 514 mg/m^3 or H_3PO_4 at 710 mg/m^3), a 15-min exposure resulted in all subjects reporting tightness in the chest, coughing, nose irritation, and difficulty in speaking. The authors stated that exposure at an average concentration of P_2O_5 at 514 mg/m^3 (H_3PO_4 at 710 mg/m^3) approaches the maximum concentration that can be tolerated for 15 min without serious effects. White and Armstrong (1935) stated that the concentration reported for the studies did not represent the maximum concentration to which the subjects were exposed, but instead represented an average of the concentration measurements taken throughout the exposure period. Thus, the maximum concentration in the chamber must have been considerably higher than the average concentration reported. For that reason, the White and Armstrong studies were not used in recommending guidance levels.

White and Armstrong (1935) conducted two additional studies in which the volunteers entered the chamber after the WP smoke concentration reached the desired level. In one study, a 2-min exposure of P_2O_5 at 588 mg/m^3 (H_3PO_4 at 812 mg/m^3) resulted in coughing, tightness in the throat, and headaches. One individual developed acute bronchitis. In the second study, six volunteers were exposed for 3.5 min at a concentration of P_2O_5 at 592 mg/m^3 (H_3PO_4 at 818 mg/m^3). The effects reported were similar to those reported for the 2-min exposure. All effects were reversible.

An accidental exposure of four females to WP smoke in a closed room for 15-20 min (concentration not reported) resulted in numerous respiratory symptoms (i.e., nose and throat irritation), edema of larynx and vocal cords, and coughing. Injury apparently extended into the bronchi. Chest X-rays revealed patchy areas of infiltration that later cleared; however, laryngitis persisted for several months (Walker et al. 1947).

Five males were exposed to WP smoke composed of phosphorus at 35 mg/m^3 and P$_2$O$_5$ at 22 mg/m^3 for 2 to 6 hr at 7-hr intervals, equivalent to H$_3$PO$_4$ at 140 mg/m^3 (total exposure time not given). Within 6 to 20 hr, all developed symptoms of weakness, dry cough, headaches, tracheobronchitis, rales, tender and enlarged liver, and evidence of leukocytosis with relative lymphocytopenia (Aizenshtadt et al. 1971, as cited in Wasti et al. 1978). Erythrocyte acetylcholinesterase was reduced by 17%, and plasma acetylcholinesterase was reduced by 35%.

No deaths were reported in humans exposed to WP smoke with H$_3$PO$_4$ at concentrations as high as 817 mg/m^3 (P$_2$O$_5$ at 592 mg/m^3) for 3 to 5 min or with H$_3$PO$_4$ at 709 mg/m^3 (P$_2$O$_5$ at 514 mg/m^3) for 15 min (White and Armstrong 1935).

There are no data on gastrointestinal, cardiovascular, musculoskeletal, hepatic, renal, dermal and ocular, immunological, neurological, reproductive, developmental, genotoxic, or carcinogenic effects from inhalation of WP smoke by humans.

WHITE PHOSPHORUS SMOKE: EFFECTS IN ANIMALS

Lethality

Lethality studies have been conducted on mice, rats, guinea pigs and goats following the inhalation of WP smoke. In mice, exposure for 1 hr resulted in mortality ranging from 5% with P$_2$O$_5$ at 110 mg/m^3 (H$_3$PO$_4$ at 150 mg/m^3) to 95% with P$_2$O$_5$ at 1,690 mg/m^3 (H$_3$PO$_4$ at 2,330 mg/m^3) (White and Armstrong 1935). The authors stated that the mice appeared to have died of mechanical obstruction to respiration, and the data probably have little or no relevance to the true toxicity of WP smoke to mice. Rats and goats appeared to be more tolerant; mortality following a 1-hr exposure ranged from 0% with P$_2$O$_5$ at 380 mg/m^3 (H$_3$PO$_4$ at

525 mg/m^3) to 100% with P$_2$O$_5$ at 4,810 mg/m^3 (H$_3$PO$_4$ at 6,200 mg/m^3) for rats and 0% with P$_2$O$_5$ at 4,810 mg/m^3 (H$_3$PO$_4$ at 6,640 mg/m^3) to 100% with P$_2$O$_5$ at 8,010 mg/m^3 (H$_3$PO$_4$ at 11,100 mg/m^3) for goats (White and Armstrong 1935). Shinn et al. (1985) estimated that the lethal concentration for 50% of the test animals (LC$_{50}$) for rats exposed to WP smoke for 1 hr was 2,500 mg/m^3 (chemical form not reported). Shinn et al. (1985) noted that the LC$_{50}$ values from four inhalation toxicity studies of rats exposed to WP smoke ranged from 1,300 to 4,800 mg/m^3 (method of measuring WP-smoke concentration not reported). The acute signs of toxicity reported were pulmonary congestion, hemorrhage, inflammation of the trachea, pneumonia, and cloudy swelling of the liver, heart, and kidney.

Rats were exposed for 60-90 min to concentrations of WP smoke with H$_3$PO$_4$ ranging from 758 to 3,030 mg/m^3. The C × T values for H$_3$PO$_4$ ranged from 45,400 to 272,000 mg·min/m^3 (Brown et al. 1980). With a 90-min exposure, the mortality of rats ranged from 0% at H$_3$PO$_4$ concentrations up to 1,200 mg/m^3 to 90% at H$_3$PO$_4$ concentrations of 3,030 mg/m^3 (Brown et al. 1980; EPA 1990). The lethal concentration for 50% of the test animals multiplied by exposure time (LCt$_{50}$) for H$_3$PO$_4$ was determined to be 141,000 mg·min/m^3 (Brown et al. 1980). In those studies, the signs of toxicity from H$_3$PO$_4$ were gasping and ataxia at 1,200 mg/m^3 (108,000 mg·min/m^3), but all animals exposed at that concentration recovered. Histopathological examination from the highest exposure group showed fibrin thrombi in heart and lungs, acute diffuse congestion, focal perivascular edema, and hemorrhage in the lungs.

With subchronic exposures (15 min per day, 5 days per week for 13 weeks) to WP smoke with H$_3$PO$_4$ at 1,740 mg/m^3, 23 of 72 rats died within 6 weeks, and a total of 29 died by the end of the 13-week study (Brown et al. 1981). There were no deaths at concentrations of H$_3$PO$_4$ of ≤884 mg/m^3.

Death occurred in all guinea pigs exposed for 30 min to WP smoke with H$_3$PO$_4$ at 716 mg/m^3 or for 60 min with H$_3$PO$_4$ at 1,200 mg/m^3 (C × T = 21,500 and 72,100 mg·min/m^3, respectively) (Brown et al. 1980). The LCt$_{50}$ estimate for H$_3$PO$_4$ was 7,980 mg·min/m^3, with a 95% confidence limit for H$_3$PO$_4$ of 7,124 to 8,943 mg·min/m^3 (Brown et al. 1980). Respiratory distress was evident in animals exposed to WP smoke with H$_3$PO$_4$ at a C × T > 8,120 mg·min/m^3. Studies that measured pulmonary resistance in guinea pigs exposed at 5,760 and 7,920 mg·min/m^3 did not reveal any difference from controls (Brown et al. 1980).

Respiratory Effects

Signs of respiratory-tract irritation (slight-to-intense congestion, edema, and hemorrhage) were observed in the lungs of mice, rats, and goats following inhalation exposure to WP smoke (White and Armstrong 1935; Brown et al. 1980). A 60-min exposure of mice, rats, and goats produced clear signs of irritation to H_3PO_4 at concentrations of 152, 525, 745 mg/m^3, respectively (White and Armstrong 1935). Longer-term exposures (15 min per day, 5 days per week for 6 or 13 weeks) of rats with H_3PO_4 at 884 mg/m^3 resulted in slight laryngitis and tracheitis (Brown et al. 1981). A similar exposure, but at higher concentrations (H_3PO_4 at 1,742 mg/m^3), resulted in wheezing, dyspnea, moderate-to-severe laryngitis and tracheitis, and interstitial pneumonia (White and Armstrong 1935; Brown et al. 1981).

Hepatic Effects

A slight clouding and swelling was observed in the livers of rats exposed for 1 hr to WP smoke with H_3PO_4 at $\geq 1,615$ mg/m^3 (White and Armstrong 1935) or at 3,027 mg/m^3 for 90 min (Brown et al. 1980). Those effects were also seen in mice and goats exposed for 1 hr at H_3PO_4 concentrations of 649 and 10,104 mg/m^3, respectively (White and Armstrong 1935). No such hepatic effects were reported in 6 of 10 guinea pigs that died from exposure for as long as 10 min to H_3PO_4 at 984 mg/m^3 (Brown et al. 1980).

Renal Effects

A slight clouding and swelling in the kidneys of rats, mice, and goats were reported following a 1-hr exposure to WP smoke with H_3PO_4 at $\geq 1,615$, 649, or 10,104 mg/m^3, respectively (White and Armstrong 1935). No renal effects were observed in rats exposed for 90 min to WP smoke with H_3PO_4 at 3,030 mg/m^3 or guinea pigs exposed for 10 min at 984 mg/m^3, respectively (Brown et al. 1980). Longer-term exposures (15 min per day, 5 day per week for 13 weeks) to WP smoke with H_3PO_4 at concentrations as high as 1,742 mg/m^3 failed to produce any significant gross or histological changes in the kidneys (Brown et al. 1981).

Reproductive and Developmental Effects

In a teratology study, pregnant female rats were exposed by inhalation to WP smoke with H_3PO_4 at concentrations of 884 and 1,742 mg/m^3 for 15 min per day for 10 days on gestation days 6-15, and fetuses were collected and observed on gestation day 20 (Starke et al. 1982). No significant maternal or developmental effects were observed in any of the exposure groups, except for an increase in certain types of visceral anomalies. In particular, the incidence of ectopic testes increased in the high-dose group compared with the control group (three in the high-dose group versus zero in the control group). Nine cases of reversed ductus arteriosus also occurred in the high-dose group compared with none in the control group. Although the authors discounted reversed ductus arteriosus as being a minor effect, the condition is suggestive of reversal of the great vessels of the heart, which is a serious cardiovascular defect.

In a dominant lethal mutation study, male rats were exposed by inhalation to WP smoke H_3PO_4 at concentrations of 884 and 1,742 mg/m^3 for 15 min per day, 5 days per week for 10 weeks (Starke et al. 1982). Four out of 18 exposed at a concentration of 1,742 mg/m^3 died during the exposure period; none out of 18 exposed at a concentration of 884 mg/m^3 died during the exposure period. After exposure, the males were mated to unexposed females. Females mated to males exposed at a concentration of 884 mg/m^3 had significantly more resorptions, but that was not a dose-related effect. No other significant effects were seen in this study. A limitation in the interpretation of this study is that the data were analyzed using the number of offspring as the experimental group size instead of using the number of males exposed to WP smoke.

In a single-generation study, F_0-generation male and female rats were exposed by inhalation to WP smoke with H_3PO_4 at concentrations of 884 and 1,742 mg/m^3 for 15 min per day, 5 days per week for 10 weeks (males) or 3 weeks (females) before mating (Starke et al. 1982). Females continued to be exposed through gestation and lactation (until day 21). F_1-generation offspring were weighed 1, 4, 7, 14, and 21 days after birth, then sacrificed and examined for gross external and visceral abnormalities on postnatal day 21. The number of pups per litter was not significantly affected, and no abnormalities were seen in a subset of pups examined. However, offspring weights were significantly less in rats exposed at 1,742 mg/m^3 than in rats exposed at 884 mg/m^3 and in the controls. Additionally, survivability, viability, and lactation indices were also significantly affected in the high-dose group. The effects were severe and were likely due to exposure to WP smoke. A limitation of the study is that the data appear to have been analyzed using the number of pups examined; the effect of variability between mated males or litters was not taken into account.

Other End Points

No significant changes in erythrocyte, hematocrit, hemoglobin, or total and differential leukocyte levels were observed in rats exposed for 90 min to WP smoke with H_3PO_4 at 3,027 mg/m^3 or in guinea pigs exposed for 10 min with H_3PO_4 at 984 mg/m^3. No hematological effects were reported in rats exposed for 13 weeks with H_3PO_4 at 1,742 mg/m^3 (Brown et al. 1981).

Rats exposed for 13 weeks (15 min per day, 5 days per week) to WP smoke with H_3PO_4 as high as 1,742 mg/m^3 showed no effects on the skin or eye or any histological alterations in the brain, heart, or gastrointestinal tract (Brown et al. 1981). No studies were identified regarding immunological, muscular-skeletal, genotoxic, or carcinogenic effects associated with the inhalation of WP smoke.

SUMMARY OF TOXICITY DATA

Tables 2-2 and 2-3 summarize the toxicological effects in animals and humans associated with exposure to WP-smoke inhalation.

PREVIOUS RECOMMENDED EXPOSURE LIMITS

Although no exposure limits have been established for WP smoke, limits have been recommended for white (yellow) phosphorus particles. WP concentrations in workplace air are regulated by the Occupational Safety and Health Administration (OSHA), and recommendations for safe levels have been made by the National Institute of Occupational Safety and Health (NIOSH) and the American Conference of Governmental Industrial Hygienists (ACGIH) (Table 2-4).

TABLE 2-2 Acute Lethality of White Phosphorus Smoke (Expressed as H_3PO_4) via Inhalation Exposure

Species	Exposure Duration	Exposure Concentration (mg/m^3)	C × T (mg·min/m^3)	End Point and Comments	Reference
Mouse	60 min	150	9,100	5% mortality	White and Armstrong 1935[a]
		800	48,000	50%	
		1,700	102,000	65%	
		2,330	139,500	95%	
Rat	60 min	525	31,500	0% mortality	White and Armstrong 1935[a]
		1,100	66,000	30%	
		3,450	210,000	50%	
		6,200	370,000	100%	
Rat	60 min	758	45,480	0% mortality	Brown et al. 1980[b]
		1,794	107,640	20%	
		1,943	116,580	50%	
		2,090	125,400	60%	
	90 min	1,200	45,500	0%	
		3,030	272,000	90%	
				LCt_{50} = 141,000 mg·min/m^3; 95% confidence interval = 115,800-172,100 mg·min/m^3	
Rat	15 min/d, 5 d/wk, for 13 wk	289		0% mortality	Brown et al. 1980[b]
		884		0%	
		1742		40%	

Species	Duration			Result	Reference
Rat	Gestation d 6-15 (15 min/d)	1,742	—	20% mortality (dams)	Starke et al. 1982[b]
Guinea pig	5 min	164	820	0% mortality	Brown et al. 1980[b]
	10 min	138	1,380	0%	
		812	8,120	60%	
		984	9,840	90%	
	30 min	192	5,760	0%	
		264	7,920	20%	
		677	20,310	90%	
		716	21,500	100%	
	60 min	1200	72,100	100%	
				LCt_{50} = 7,980 mg·min/m^3; 95% confidence interval =7,124-8,943 mg·min/m^3	
Goat	60 min	6,640	400,000	0% mortality	White and Armstrong 1935[a]
		10,000	601,000	50%	
		11,100	664,000	100%	

[a]Originally reported as P_2O_5 equivalents; multiplied by 1.381 to estimate H_3PO_4 equivalents.
[b]Reported values multiplied by 1.5 to correct for acid dissociation at pH of 9.6 to estimate H_3PO_4 equivalents.
Abbreviations: C × T, the product of exposure concentration and time; LCt_{50}, lethal concentration for 50% of the test animals multiplied by exposure time.

TABLE 2-3 Nonlethal Effects of White Phosphorus Smoke (Expressed as H_3PO_4) via Inhalation Exposure

Species	Exposure Duration	Exposure Concentration (mg/m^3)	NOAEL (mg/m^3)	LOAEL (mg/m^3)	End Point and Comments	Reference
				Respiratory Effects		
Human	Not specified	87-1,770	—	700 1,000	Coughing, respiratory irritation Irritation intolerable	Wasti et al. 1978; EPA 1990; ATSDR 1997
Human	2 min	812	—	812	Coughing, tightness in throat, and headaches; one developed acute bronchitis	White and Armstrong 1935
Human	3.5 min	818	—	818	Same symptoms as 812 mg/m^3 for 2-min exposure.	White and Armstrong 1935
Human	5 min	259-710 (average)	—	259 (average)	50% indicated sore throat, irritation, coughing, and congestion	White and Armstrong 1935
Human	15 min	259-710 (average)	—	710 (average)	All experienced tightness of chest, difficulty speaking; suggested maximum concentration without serious effects	White and Armstrong 1935
Human	15-20 min	Not specified	—	—	Choking, tightness in chest, rales, sore throat, sputum production	Walker et al. 1947; EPA 1990
Rat	60 min	758-3,030	—	758 3,030	Respiratory distress Intense congestion, edema, hemorrhages	Brown et al. 1980[a]

Species	Duration	Dose range			Effect	Reference
Rat	60 min	525-6,640	—	525	Pulmonary congestion, hemorrhages, respiratory distress, unmistakable signs of irritation	White and Armstrong 1935[b]
Rat	90 min	1,200	—	1,200	Grasping, ataxia, respiratory distress	Brown et al. 1980[a]
Guinea pig	10 min	138, 812, 984	138	—	No respiratory effects	Brown et al. 1980[a]
Mouse	60 min	152-2,330	—	152	Unmistakable signs of irritation, congestion, difficulty breathing	White and Armstrong 1935[b]
Goat	60 min	745-15,840	—	745	Unmistakable signs of irritation, inflammation, pneumonia	White and Armstrong 1935[b]
Rat	15 min/d, 5 d/wk, for 13 wk	280, 884, 1,742	280	884 1,742	Slight tracheitis and laryngitis; Moderate-to-severe tracheitis and laryngitis	Brown et al. 1981[a]
Hepatic Effects						
Rat	60 min	525-6,640	—	$\geq$1,615	Slight clouding and swelling of liver in 1 rat; seen more consistently $\geq$3,494 mg/m^3	White and Armstrong 1935[b]
Rat	90 min	3,030	—	3,030	Liver congestion	Brown et al. 1980[a]
Rat	15 min/d, 5 d/wk, 13 wk	289-1,742	1,742	—	No gross or histological changes	Brown et al. 1981[a]
Guinea pig	10 min	138, 812, 984	984[c]	—	No gross or histological changes	Brown et al. 1980[a]

TABLE 2-3 (Continued)

Species	Exposure Duration	Exposure Concentration (mg/m³)	NOAEL (mg/m³)	LOAEL (mg/m³)	End Point and Comments	Reference
Guinea pig	30 min	192	192	—	No gross or histological changes	Brown et al. 1980[a]
Mouse	60 min	152-2,330	—	649	Slight swelling and clouding of liver	White and Armstrong 1935[b]
Goat	60 min	10,104	—	10,104	Slight swelling and clouding of liver	White and Armstrong 1935[b]
Renal Effects						
Rat	60 min	525-6,640	—	≥1,615	Slight swelling and clouding of kidney	White and Armstrong 1935[b]
Rat	90 min	3,030	—	—	No gross or histological changes	Brown et al. 1980[a]
Rat	15 min/d, 5 d/wk, 13 wk	289, 884, 1,742	1,742	—	No gross or histological changes	Brown et al. 1981[a]
Guinea pig	10 min	138, 812, 984	984[c]	—	No gross or histological changes	Brown et al. 1980[a]
Guinea pig	30 min	192	192	—	No gross or histological changes	Brown et al. 1980[a]
Mouse	60 min	152-2,330	428	649	Slight swelling and clouding of kidney	White and Armstrong 1935[b]

Species	Duration				Effects	Reference
Goat	60 min	10, 104	—	10, 104	Slight swelling and clouding of kidney	White and Armstrong 1935[b]
Reproductive and Developmental Effects						
Rat (male)	15 min/d, 5 d/wk, 10 wk	884, 1,742	884	—	4 of 18 rats died in the high-dose group; no deaths in the low-dose group; no other significant effects observed	Starke et al. 1982
Rat (female)	15 min/d, gestation d 6-15	884, 1,742	884	—	Increase in incidence of ectopic testes and reversed ductus arteriosus in the high-dose group; no effects in the low-dose group	Starke et al. 1982
Rat (male and female)	Males: 15 min/d, 5 d/wk, 10 wk before mating; Females: 15 min/d, 5 d/wk, 3 wk before mating and through gestation and lactation to d 21	884, 1,742	884	—	Offspring body weight, survivability, and viability reduced in high-dose group; no significant effect on number of pups per litter and no abnormalities observed in either exposure group	Starke et al. 1982

[a]Reported value multiplied by 1.5 to correct for acid dissociation at pH of 9.6 to estimate H_3PO_4 equivalents.
[b]Originally reported as P_2O_5 equivalents; multiplied by 1.381 to estimate H_3PO_4 equivalents.
[c]Highest NOAEL for a given species and study when exposure expressed as CxT.
Abbreviations: NOAEL, no-oberserved-adverse-effect level; LOAEL, lowest-observed-adverse-effect level; C X T, the product of exposure concentration and time.

TABLE 2-4 Existing Exposure Limits for White (Yellow) Phosphorus Particles

Agency	Description	Level (mg/m^3)	Reference
OSHA	PEL-TWA	0.1	U.S. Dept. of Labor 1997
ACGIH	TLV-TWA	0.1	ACGIH 1991
NIOSH	REL-TWA	0.1	NIOSH 1996

Abbreviations: OSHA, Occupational Safety and Health Administration; ACGIH, American Conference of Governmental Industrial Hygienists; NIOSH, National Institute of Occupational Safety and Health; PEL-TWA, permissible exposure limit–time-weighted average; TLV-TWA, Threshold Limit Value–time-weighted average; REL-TWA, recommended exposure limit–time-weighted average.

EPA (1998) has listed WP as a hazardous air pollutant and has classified it as a Group D carcinogen (inadequate evidence of carcinogenicity). Various states also have established acceptable ambient concentration guidelines or standards for different exposure durations (see examples in Table 2-5).

SUBCOMMITTEE EVALUATION AND RECOMMENDATIONS

On the basis of the available toxicity information, the subcommittee recommended exposure guidance levels for military personnel exposed during an emergency release and during regular training exercises and for consideration at training-facility boundaries to protect nearby communities from an acute exposure or repeated releases of WP smoke.

MILITARY EXPOSURES

Emergency Exposure Guidance Levels (EEGLs)

In recommending the EEGLs for WP smoke, the most sensitive response to short-term exposure is respiratory irritation and distress. Animal studies indicate that such an effect becomes evident in goats and rats following a 1-hr exposure to WP smoke with H_3PO_4 at 745 and 525 mg/m^3, respectively (P_2O_5 at 540 and 380 mg/m^3, respectively). Human volunteers exposed for 3.5 min to WP smoke with H_3PO_4 at 818 mg/m^3 (P_2O_5 at 592 mg/m^3) reported respiratory irritation, tightness of chest, cough, and difficulty in breathing (White and Armstrong 1935). In those

TABLE 2-5 Selected State Guidelines for White Phosphorus (Expressed as Elemental Phosphorus)

Texas	1.7 $\mu g/m^3$ (24 hr)
Florida	1.0 $\mu g/m^3$ (8 hr)
Connecticut	2.0 $\mu g/m^3$ (8 hr)
Oklahoma	1.0 $\mu g/m^3$ (30 min)

Source: ATSDR 1997.

cases, the subjects refused to be exposed to higher concentrations and thought it would be impossible, without more serious effects, to perform any physical exercise or labor at that concentration.

The subcommittee considered both human and animal data in recommending the EEGLs. Several EEGLs were estimated on the basis of data from humans and animals, as shown in Table 2-6. The lowest-observed-adverse-effect level (LOAEL) identified in mice of 152 mg/m^3 was not included in the subcommittee's evaluation because the mice apparently were sensitive to mechanical obstruction to respiration and the data probably had little or no relevance to the true toxicity of WP smoke to mice (White and Armstrong 1935). Therefore, using the animal LOAEL identified in rats of 525 mg/m^3 (White and Armstrong 1935), an uncertainty factor of 10 was used to extrapolate from a LOAEL to a no-observed-adverse-effect level (NOAEL), and an additional factor of 10 was used to extrapolate from animal to human. Assuming that Haber's rule (that is, the product of exposure concentration and time is a constant, $C \times T = k$) applied, the estimated EEGLs were calculated to be 21, 5, and 1 mg/m^3 for WP smoke expressed as H_3PO_4 for 15 min, 1 hr, and 6 hr, respectively. The estimated EEGLs derived from human data were calculated to be 19, 5, and 0.8 mg/m^3 for WP smoke expressed as H_3PO_4 for 15 min, 1 hr, and 6 hr, respectively. The subcommittee's decision to use the data from the White and Armstrong (1935) study to recommend EEGLs is supported by a more recent inhalation toxicity study (Brown et al. 1980). In that more recent study, rats were exposed for 90 minutes at 707 mg/m^3 and showed signs of gasping and became ataxic, but recovered. Because human data are available and the animal data are consistent with the human data, the subcommittee recommends using EEGLs derived from human data.

The subcommittee recognizes that these EEGLs are lower than those

TABLE 2-6 Estimated EEGLs from Human and Animal Data for White Phosphorus Smoke (Expressed as H_3PO_4)

Species	LOAEL Exposure Duration	LOAEL mg/m^3	Uncertainty Factors LOAEL to NOAEL	Uncertainty Factors Animals to Humans	Estimated EEGLs (mg/m^3) Duration 15 min	1 hr	6 hr
Human	3.5 min	818	10	—	19	5	0.8
Rat	1 hr	525	10	10	21	5	1
Goat	1 hr	745	10	10	30	7	1
Recommended EEGLs expressed as H_3PO_4					19	5	0.8
Recommended EEGLs expressed as P_2O_5					14	4	0.6
Recommended EEGLs expressed as phosphorus					6	1.6	0.3

Abbreviations: EEGLs, emergency exposure guidance levels; NOAEL, no-observed-adverse-effect level; LOAEL, lowest-observed-adverse-effect level.

recommended by this subcommittee for RP-BR (NRC 1997), even though the final combustion products for both smokes are expected to be phosphoric acid. However, the human and animal data indicate that WP smoke appears to produce respiratory irritation at lower concentrations than does RP-BR smoke. For example, the rats exposed for 1 hr to RP-BR showed signs of respiratory irritation with H_3PO_4 at 1,692 mg/m^3, but none died (Weimer et al. 1977). Rats exposed to WP smoke, with H_3PO_4 at 1,794 mg/m^3, a concentration similar to that of RP-BR, had a 20% mortality during the 1-hr exposure (Brown et al. 1980). The difference might result from the presence of some uncombusted WP in the WP smoke. That would contribute to the smoke's toxicity.

A similar difference in sensitivity to the two phosphorus smokes can be observed when comparing the human data on WP and RP-BR smokes. For example, Mitchell and Burrows (1990) stated that acute exposure to RP-BR smoke at 1,000 mg/m^3 (chemical form not reported) would be intolerable and that 700 mg/m^3 (chemical form not reported) is the highest tolerable concentration. In contrast, White and Armstrong (1935) stated that human volunteers exposed to WP smoke with P_2O_5 at 592 mg/m^3 (H_3PO_4 at 818 mg/m^3) said that was the limit of their tolerance.

Repeated Exposure Guidance Level (REGL)

Although there is an existing ACGIH Threshold Limit Value–time-weighted average (TLV-TWA) for WP particles (0.1 mg/m^3), that value does not seem appropriate for WP smoke. The short-term exposure data for WP smoke suggest that the REGL (8 hr per day, 5 days per week) could be higher than the TLV-TWA of 0.1 mg/m^3 for WP alone but should be lower than the TLV-TWA of 1 mg/m^3 for phosphoric acid. Rats exposed for 15 min per day, 5 days per week for 13 weeks showed moderate laryngitis and tracheitis with H_3PO_4 at 1,400 mg/m^3, and slight laryngitis and tracheitis with H_3PO_4 at 690 mg/m^3, and no such effects were reported with H_3PO_4 at 280 mg/m^3. Because that result identifies a no-effect exposure concentration, the subcommittee recommends that it be used to establish a REGL. An uncertainty factor of 10 is used to extrapolate the animal data to humans. Dividing 28 mg/m^3 by 32 to estimate a value for 8 hr from a 15-min exposure using Haber's rule yields a value of 0.9 mg/m^3. Applying an uncertainty factor of 10 to extrapolate from subchronic to chronic exposure yields a value of 0.09 mg/m^3. Therefore, the subcommittee recommends the REGL for WP (expressed as H_3PO_4) to be 0.09 mg/m^3 for 8 hr per day, 5 days per week.

PUBLIC EXPOSURES

Short-Term Public Emergency Guidance Levels (SPEGLs)

Assuming that the general population includes a variety of susceptible individuals, an additional uncertainty factor of 10 is appropriate to extrapolate from an EEGL for military personnel to a level protective of the general public. Thus, the SPEGLs for a single emergency exposure to WP smoke expressed as H_3PO_4 are 1.9, 0.5, and 0.08 mg/m^3 for exposures of 15 min, 1 hr, and 6 hr, respectively.

SUMMARY OF SUBCOMMITTEE RECOMMENDATIONS

The subcommittee's recommendations for exposure to WP smoke for military personnel and for the public (i.e., the boundaries of military-training facilities) are summarized in Tables 2-7 and 2-8.

RESEARCH NEEDS

The subcommittee recognizes the need for further research to better understand the potential toxicity associated with inhaling WP smoke. Research in the following areas would provide better insight into possible health effects of exposure to WP smoke and help to determine, with greater confidence, a guidance level that is not overly conservative but is scientifically defensible in ensuring minimal risk to the exposed population. The subcommittee recommends that the following studies be conducted:

- Studies designed to measure the absorption, distribution, and metabolism of WP smoke would be of help in assessing human risk.
- Only a limited number of biological and biochemical end points have been studied in relation to inhalation of WP smoke. Inhalation

TABLE 2-7 EEGLs and REGL for White Phosphorus Smoke (Expressed as H_3PO_4) for Military Personnel

Exposure Guideline	Exposure Duration	Exposure Guidance Level (mg/m³)
EEGL	15 min	19
	1 hr	5
	6 hr	0.8
REGL	8 hr/d, 5 d/wk	0.09

Abbreviations: EEGL, emergency exposure guidance level; REGL, repeated exposure guidance level.

TABLE 2-8 SPEGLs and RPEGL for White Phosphorus Smoke (Expressed as H_3PO_4) at the Boundaries of Military-Training Facilities

Exposure Guideline	Exposure Duration	Exposure Guidance Level (mg/m³)
SPEGL	15 min	1.9
	1 hr	0.5
	6 hr	0.08
RPEGL	8 hr/d, 5 d/wk	0.009

Abbreviations: SPEGLs, short-term public emergency guidance level; RPEGL, repeated public exposure guidance level.

studies are needed to evaluate possible neurobehavioral, immunological, reproductive and developmental, and carcinogenic effects of WP smoke.

• All studies should include improved characterization of the composition and particle properties of the WP smoke used and actual exposure concentrations.

• Long-term chronic inhalation studies would be of value to identify target organs of toxicity and possible long-term health effects.

REFERENCES

ACGIH (American Conference of Governmental Industrial Hygienists). 1991. Documentation of the Threshold Limit Values and Biological Exposure Indices, 6th Ed. American Conference of Governmental Industrial Hygienists, Cincinnati, Ohio.

Aizenshtadt, V.S., S.M. Nerubai, and Y.I. Voronin. 1971. Clinical aspects of acute poisoning by vapors of phosphorus and its oxides under industrial conditions [in Russian]. Gig. Tr. Prof. Zabol. 15(10):48-49.

ATSDR (Agency for Toxic Substances and Disease Registry). 1997. Toxicological Profile for White Phosphorus. U.S. Department of Health and Human Services, Agency for Toxic Substances and Disease Registry, Atlanta, Ga.

Brazell, R.S., J.H. Moneyhun, and R.W. Holmberg. 1984. Chemical Characterization and Toxicological Evaluation of Airborne Mixtures. ORNL/TM-9571. AD-A153824. Oak Ridge National Laboratory, Oak Ridge, Tenn.

Brown, B.J., J.T. Weimer, G.E. Affleck, R.L. Ferrand, D.H. Heitkamp, and F.K. Lee. 1980. The Acute Effects of Single Exposures to White Phosphorus Smoke in Rats and Guinea Pigs. ARCSL-TR-80013. AD-B051836L. Chemical Systems Laboratory, U.S. Army Armament, Munitions and Chemical Command, Aberdeen Proving Ground, Edgewood, Md.

Brown, B.J., G.E. Affleck, E.G. Cummings. R.L. Farrand, W.C. Starke, J.T. Weimer, M.S. Ghumman, and R.J. Pellerin. 1981. The Subchronic Effects of Repeated Exposure to White Phosphorus/Felt Screening Smokes in Rats. ARCSL-TR-80068. AD-B05848L. Chemical Systems Laboratory, U.S. Army Armament, Munitions and Chemical Command, Aberdeen Proving Ground, Edgewood, Md.

Cullumbine, H. 1944. The Burning Power and Harassing Effects of White Phosphorus. Porton Report No. 2604. Military Intelligence Division, Great Britain.

Ellis, H.V. III, J.R. Hodgson, S.W. Hwang, L.M. Halpap, D.O. Hetton, B.S. Anderson, D.L. Van Goethem, and C.C. Lee. 1978. Mammalian Toxicity of Munitions Compounds. Phase 1: Acute Oral Toxicity, Primary Skin and Eye Irritation, Dermal Sensitization, Disposition and Metabolism, and Ames Tests of Additional Compounds. DAMD17-74-C-4073. Prepared by Midwest Research Institute, Kansas City, Mo., for the U.S. Army Medical Research and Development Command, Frederick, Md.

EPA (U.S. Environmental Protection Agency). 1990. Summary Review of Health Effects Associated with Elemental and Inorganic Phosphorus Compounds: Health Issue Assessment. EPA 600/8-89/072. U.S. Environmental Protection Agency, Environmental Criteria and Assessment Office, Research Triangle Park, N.C.

EPA (U.S. Environmental Protection Agency). 1998. White phosphorus. In Integrated Risk Information System (IRIS). Online. Entry last revised Feb. 1, 1993. Available: http://www.epa.gov/iris/subst/0460.htm#II. U.S. Environmental Protection Agency, National Center for Environmental Assessment, Cincinnati, Ohio.

Heimann, H. 1946. Chronic phosphorus poisoning. J. Ind. Hyg. Toxicol. 28:142-150.

Hughes, J.P., R. Baron, D.H. Buckland, M.A. Cooke, J.D. Craig, D.P. Duffield,

A.W. Grosart, P.W. Parkes, and A. Porter. 1962. Phosphorus necrosis of the jaw: A present day study. Br. J. Ind. Med. 19:83-99.

Katz, S., A. Snelson, R. Butler, W. Bock, N. Rajendran, and S. Relwani. 1981. Physical and Chemical Characterization of Military Smokes. DAMD17-78-C-8085. Prepared by IIT Research Institute, Chicago, for the U.S. Army Medical Research and Development Command, Frederick, Md.

Lee, C.-C., J.V. Dilley, J.R. Hodgson, D.O. Helton, W.J. Wiegand, D.N. Roberts, B.S. Andersen, L.M. Halfpap, L.D. Durtz, and N. West. 1975. Mammalian Toxicity of Munition Compounds: Phase I. Acute Oral Toxicity, Primary Skin and Eye Irritation, Dermal Sensitization and Disposition and Metabolism. U.S. Army Medical Research and Development Command, Fort Detrick, Md.

Lide, D.R., ed. 1991. P. 8-41 in CRC Handbook of Chemistry and Physics. 72nd Ed. Boca Raton, Fla.: CRC Press.

Maruo, T. 1955. Experimental Study on poisoning due to gas of phosphor. Part 1. The hemogram of rabbits [in Japanese]. Fukuoka Acta Med. 46:604-615.

Mitchell, W.R., and E.P. Burrows. 1990. Assessment of Red Phosphorus in the Environment. Tech. Rep. 9005. AD-A221704. U.S. Army Biomedical Research and Development Laboratory, Frederick, Md.

NIOSH (National Institute for Occupational Safety and Health). 1992. Recommendations for Occupational Safety and Health: Compendium of Policy Documents and Statements. DHHS (NIOSH) Publ. No. 92-100. National Institute for Occupational Safety and Health, Cincinnati, Ohio.

NRC (National Research Council). 1997. Toxicity of Military Smokes and Obscurants, Vol. 1. Washington, D.C.: National Academy Press.

Ruzuddinov, S.R., and M. Rys-Uly. 1986. Morphology of the oral mucosa in rats exposed to increased phorphorus levels. Bull. Exp. Biol. Med. 101:109-112.

Shinn, J.H., S.A. Martins, P.L. Cederwall, and L.B. Gratt. 1985. Smokes and Obscurants: A Health and Environmental Effects Data Base Assessment. Phase I Report. UCID-20931. U.S. Army Medical Research and Development Command, Fort Detrick, Md.

Simon, F.A., and L.K. Pickering. 1976. Acute yellow phosphorus poisoning: Smoking stool syndrome. JAMA 235:1343-1344.

Spanggord, R.J., R. Renwick, T.-W. Chou, R. Wilson, R.T. Podoll, T. Mill, R. Parnas, R. Platz, and D. Roberts. 1985. Environmental Fate of White Phosphorus/Felt and Red Phosphorus/Butyl Rubber Military Screening Smokes. Final Report. DAMD17-82-C-2320. AD-176922. Prepared by SRI International, Menlo Park, Calif., for U.S. Army Medical Research and Development Command, Frederick, Md.

Starke, W.C., R.J. Pellerin, and D.C. Barnett. 1982. White Phosphorus-Felt Smoke: Effects on Reproduction in the Rat. ARCSL-TR-82002. AD-A118682. Chemical Systems Laboratory, U.S. Army Armament, Munitions and Chemical Command, Aberdeen Proving Ground, Edgewood, Md.

U.S. Department of Labor. 1997. Occupational Safety and Health Standards. Air Contaminants. Code of Federal Regulations, Title 29, Part 1910, Section 1910.1000. Washington, D.C.: U.S. Government Printing Office.

Walker, J., Jr., M. Galdston, and J. Wexler. 1947. WP Casualties at Edgewood Arsenal, Maryland, 1945. Rep. 103. AD 824420. U.S. Army Armament, Munitions and Chemical Command, Aberdeen Proving Ground, Edgewood, Md.

Wasti, K., K.J.R. Abaidoo, and J.E. Villaume. 1978. A Literature Review-Problem Definition Studies on Selected Toxic Chemicals. Vol. 2. Occupational Health and Safety Aspects of Phosphorus Smoke Compounds. AD-A056019. U.S. Army Armament, Munitions and Chemical Command, Aberdeen Proving Ground, Edgewood, Md.

Weimer, J.T., G. Affleck, J. Preston, J. Lucey, J. Manthei, and F. Lee. 1977. The Acute Effects of Single Exposures to United Kingdom Red Phosphorus Screening Smoke in Rats, Guinea Pigs, Rabbits, and Dogs. ARCSL-TR-77052. Chemical Systems Laboratory, U.S. Army Armament, Munitions and Chemical Command, Aberdeen Proving Ground, Edgewood, Md.

White, S.A., and G.C. Armstrong. 1935. White Phosphorus Smoke: Its Irritating Concentration for Man and Its Toxicity for Small Animals for One-hour Exposures. E.A.T.R. 190, Project A 5.2-1. War Department, Chemical Warfare Service, Edgewood Arsenal, Aberdeen Proving Ground, Edgewood, Md.

Yon, R.L., R.S. Wentsel, and J.M. Bane. 1983. Red, White and Plasticized White Phosphorus, Vol. 2, p. 24, of Programmatic Life Cycle Environmental Assessment for Smoke/Obscurants. ARCSL-EA-83004. Chemical Systems Laboratory, Chemical Research and Development Center, U.S. Army Armament, Munitions and Chemical Command, Aberdeen Proving Ground, Edgewood, Md.

3
Brass Smoke

BACKGROUND INFORMATION

MILITARY APPLICATIONS

BRASS FLAKES are used in smoke grenades to block detection of infrared waves for thermal imaging systems. Currently, the U.S. Army uses a product called EA 5763 that contains brass flakes. Other products containing brass flakes developed by the military include EA 5763D and EA 5769, but those are not in use.

PHYSICAL AND CHEMICAL PROPERTIES

The brass flakes in EA 5763 (composed of 70% copper (Cu) and 30% zinc (Zn)) generally have a mass median aerodynamic diameter (MMAD) of 2.1 to 2.3 micrometers (μm). Sometimes the flakes are called "dust" or "powder," but the material is really an irregular flake with a diameter of approximately 1.7 μm and a thickness of 0.08 to 0.32 μm. To facilitate manufacture, the flakes are coated with palmitic or stearic acid or both (Wentsel 1986).

Analysis of flakes of EA 5763 generally reveals trace amounts of aluminum (0.2%), antimony (0.1%), and lead (0.1%), but other metals might be present that are below detectable limits (0.25 microgram per milliliter (μg/mL) by atomic absorption spectroscopy (Thomson et al. 1985).

Occurrence and Use

In addition to the use of brass flakes by the military, the brass-foundry industry produces small particles of brass that can be inhaled. Workers in this industry, particularly workers who polish brass pots, can be exposed to brass dust at concentrations as high as 0.19 mg/m^3 (combined concentrations of Cu and Zn) (Rastogi 1992a,b).

COMBUSTION PRODUCTS

The smoke from the M76 grenade has the same chemical and physical properties as the bulk material. Explosion of the grenades simply releases the brass flakes that serve to block detection of infrared waves.

TOXICOKINETICS

ABSORPTION AND DISTRIBUTION

No studies have been published on the distribution of inhaled EA 5763. However, evaluations of absorption and distribution have been conducted on a similar material, EA 5769, that the Army investigated at one time but subsequently dropped (Muse 1983). EA 5769 flakes that were used for the study had a geometric mean diameter of 5.5 μm or 4.9 μm for two samples taken from the inhalation chamber (Feeney et al. 1983), whereas the EA 5763 flakes generally have a MMAD size of about 2 μm. The overall conclusions about the characteristics of EA 5769 should be applicable to EA 5763. A principal difference in the deposition between these two particles is that EA 5763 deposition should be greater in the lower respiratory tract because of the smaller size of those brass flakes.

Fischer 344 (F344) rats (four males and four females) were exposed to EA 5769 for 15 min at a concentration of 1,000 milligrams per cubic meter (mg/m^3) (Muse 1983). In an additional group, four males and four females were exposed at 2,000 mg/m^3 for 30 min, but that dosage was lethal to all the animals within 24 hr. For the 1,000-mg/m^3 exposure group, tissue concentrations of Cu were determined immediately after exposure and 5, 7, and 14 days after exposure (Muse 1983). Immediate examination showed high concentrations of Cu (150 to 452 μg of Cu per gram (g) of dry weight) in the larynx, trachea, esophagus, stomach, upper right lung, lower right lung, upper left lung, and lower left lung. Of the tissues with initially high Cu concentrations, only the lung had high concentrations of Cu 5 days after exposure. By day 5, the liver and kidney also showed increased Cu concentrations (23 and 38 µg/g of dry weight, respectively) until day 7 (the liver) and day 14 (the kidney). By day 14, no tissues had increased concentrations of Cu.

Snipes et al. (1988) exposed rats to EA 5763 brass flakes at concentrations of 0.32, 1.0, 3.2, or 10 mg/m^3 for 1.5 hr per day, 4 days per week for 13 weeks. The amount of Cu and Zn retained in the lungs was measured at the end of the exposure and 4 weeks later for the three highest-dose groups. The amount of the two metals combined in the lung at the end of the 13-week exposure was approximately 23 µg per animal; 4 weeks later it dropped to approximately 19 µg. The similarity in the residual amount of Cu and Zn retained in the lungs of animals from the three highest-dose groups, despite an order of magnitude difference in exposure concentrations, indicates a limit to the accumulation of the brass in the lungs. The data also indicate that the clearance time for low residual concentrations after the last dose was greater than 4 weeks.

METABOLISM AND EXCRETION

Feces was the primary route of elimination of Cu from an inhaled dose of EA 5769 brass flakes in a study of F344 rats (Muse 1983). Urinary content of Cu within the first day of collection after a 15-min exposure to EA 5769 at 1 mg/L was 190 mg of Cu per kilogram (kg) of dry weight (freeze-dried), and fecal content at that time was 1,165 mg/kg of dry weight. Urinary concentrations dropped to control values by day 5, and fecal concentrations never dropped below 150 mg/kg of dry weight up to 13 days after exposure. No data were presented concerning the metabolism of the brass flakes.

TOXICITY SUMMARY

EFFECTS IN HUMANS

Human data are not available on the effects of brass flakes (the form to which military personnel might be exposed from launching brass grenades). However, the effects of chronic exposures to brass fumes from the brass industry in India have been published (Rastogi et al. 1991, 1992a,b). The particulate size of brass fumes is smaller than that of brass flakes, and therefore, fumes are likely to be more toxic than flakes.

The studies examined the prevalence of chronic bronchitis in over 580 workers in various suboccupations in the brass industry. The findings were compared with those obtained from unexposed workers. Those studies indicated that long-term exposures (10-17 years) to fumes containing brass can be associated with increased risk of chronic bronchitis, with an odds ratio risk of 2.74. In addition, there was a direct correlation between length of exposure and risk. No effects were observed in solders chronically exposed to brass fumes at 0.001 mg/m^3 (Rastogi et al. 1991), and respiratory effects were observed only in polishers chronically exposed to an average Cu-plus-Zn concentration of 0.19 mg/m^3 (Rastogi et al. 1992a,b). The workers were also exposed to other toxic metals, including lead, nickel, cadmium, manganese, and chromium. Those metals might also contribute to the respiratory effects observed in the study. The risk associated with short-term exposures to brass flakes is not known for humans.

EFFECTS IN ANIMALS

Inhalation Exposures

One-Time Exposures

Lethality. Guinea pigs are the most susceptible species to brass flakes. Exposure to EA 5763 at 100 mg/m^3 for 150 min caused respiratory failure and death in 12 out of 12 animals. Pathological findings included severe edema and bronchoconstriction (Thomson et al. 1982b). Rats are less susceptible. Groups of 12 (six males and six females) F344 rats were exposed by inhalation to brass flakes (EA 5763) at 1,000 mg/m^3 for 30 to 90 min (Feeney et al. 1983). The shortest exposure was not lethal to the rats. At exposure times greater than 50 min, at least 6 of the 12 animals died within 24 hr of the end of the exposure (Feeney et al. 1983). Groups of 12 (six males and six females) B6C3F$_1$ mice were also exposed at 1,000 mg/m^3 for durations ranging from 15 min to 300 min. The mice were less susceptible than the rats (Feeney et al. 1983). The calculated lethal concentration for 50% of the test animals multiplied by exposure time (LCt$_{50}$) for rats and mice were 53,538 mg•min/m^3 and 199,511 mg•min/m^3, respectively. Thus, acute toxicity studies show a species susceptibility that is greater in guinea pigs than rats and mice and greater in rats than mice.

Pulmonary Effects. Pathological lesions in the respiratory systems of rodents exposed to high concentrations (1,000 mg/m^3) of brass flakes were characterized largely by moderate necrosis of the turbinate epithelium, necrotizing bronchiolitis, moderate pneumonia, and alveolar histiocytic proliferation (Feeney et al. 1983). In rats, a lower concentration (100 mg/m^3 for 4 hr) produced an acute inflammatory response and pulmonary alveolar macrophage activation (Anderson et al. 1988). The macrophages isolated from exposed animals by lavage showed morphological and functional abnormalities.

Acute toxicity studies (4 hr) using F344 rats exposed at concentrations of 200, 100, 50, 10, and 1 mg/m^3 showed dose-related changes in biochemical, morphological, and functional measurements (Thomson et al. 1986). Rats examined within 24 hr of exposure showed increases in lactate dehydrogenase and protein in the lung lavage fluid. In addition, the animals showed an acute inflammatory response in the terminal airways, comprising increases in macrophages and neutrophils. The rats also showed an increased pulmonary resistance. No pathological and only minimal cytological effects were observed after a 4-hr exposure at a concentration of 1-mg/m^3.

Gastrointestinal Effects. None of the available studies reported examination of gastrointestinal tissues after inhalation of brass flakes.

Mutagenic Effects. No information was found concerning possible mutagenic effects of a single exposure to brass flakes.

Reproductive and Developmental Toxicity. No information was found concerning the reproductive or developmental toxicity of a single exposure to brass flakes by the inhalation route.

Repeated Exposures

Lethality. Exposure of F344 rats to brass flakes at a concentration of 100 mg/m^3 produced no lethality after a single 4 hr exposure (Thomson et al. 1986). That concentration was lethal to 9 of 72 rats exposed for 15 min per day, 5 days per week for 13 weeks (Thomson et al. 1982b). Lethality was observed during the second to ninth week of the exposure. Thus, repeated exposures to 100 mg/m^3 for only 15 min per day were significantly more lethal than a single exposure to the same concentration of brass flakes. Longer exposures of 150 min per day caused 23 of 72 rats to die within only 7 days; the experiment was prematurely terminated due to high lethality from repeated exposures.

Pulmonary Effects. A comparison of multiple exposures (100 mg/m^3 for 2.5 hr per day for 6 days) to a single exposure (100 mg/m^3 for 2.5 hr) showed that multiple exposures to brass flakes were significantly more toxic than single exposures (Thomson et al. 1982a). Cytological analysis, enzyme activities, and protein contents were measured in bronchoalveolar lavage fluid as indices of damage. Multiple exposures caused an immediate increase in polymorphonuclear leukocytes along with a decrease in macrophages. After 1 week in clean air, those effects were reversed, but 4 weeks were required for those measurements to return to control values. The changes were believed to be indicative of an acute inflammatory response.

The activities of enzymes, such as alkaline phosphatase, lactate dehydrogenase, and glucose-6-phosphate dehydrogenase, in the lung lavage fluid were generally not increased by a single exposure but were transiently increased for the multiple exposures (Thomson et al. 1982a). Alkaline phosphatase was markedly decreased by multiple exposures when measured from 24 hr to 2 weeks after exposure. Lactate dehydrogenase activity increased in rats exposed once or more than once and did not return to control values until 4 weeks after exposure. Protein levels in the lavage fluid of mice were much higher after multiple exposures than after a single exposure, but the single exposure did cause about a 10-fold increase in protein content.

Exposures of rats at lower concentrations of brass flakes (1 and 10 mg/m^3) for 6 hr per day, 5 days per week for 6 or 13 weeks, showed both enzymatic and cytological changes (Thomson et al. 1984). The high dose caused irreversible granulomatous pneumonia along with increased lung weight 3 months after the end of the exposure. The low dose caused only mild pulmonary lesions, which were reversible 30 days after exposure.

Rats were exposed to EA 5763 brass flakes at concentrations of 0.32, 1.0, 3.2, or 10 mg/m^3 for 1.5 hr per day, 4 days per week for 13 weeks (Snipes et al. 1988). Several end points were evaluated by bronchoalveolar lavage, immunological methods, histopathology, and respiratory function. Toxicities were observed at the three highest doses, and dose-response effects were consistently observed. At 1.0 mg/m^3, only a mild focal atrophy of olfactory epithelium was observed, and this resolved within 4 weeks of the exposure. However, the authors of this study attempted to relate human nasal exposures to rat nasal exposures. They concluded that exposure of the upper respiratory tract of rats would be approximately 20-fold higher than that exposure of humans at the same concentration of brass flakes. Thus, humans would be much less susceptible than rats to nasal toxicity from an exposure to brass flakes at a concentration of 1.0 mg/m^3.

Other Systemic Effects. Exposure of rats to brass flakes at 100 mg/m^3 for up to 13 weeks was followed by histopathological examination (Thomson 1982b) of all major organs (heart, lung, liver, spleen, kidney, brain, eye, trachea, nasal turbinate, adrenal, stomach, urinary bladder, pancreas, thyroid, esophagus, duodenum, colon, lymph node, thymus, testes, epididymus, ovary, uterus, bone marrow, and skin). Damage was confined solely to the respiratory system.

Carcinogenic and Mutagenic Effects. No information is available concerning the potential for carcinogenic effects of brass flakes. Male fruit flies (*Drosophila melanogaster*) were exposed for 72 hr at concentrations of 1.0% to 25% brass flakes mixed in the food (Manthei et al. 1983). After exposed males were mated with virgin unexposed flies, the offspring were back-crossed, and third generation flies did not show any observable mutations.

Reproductive and Developmental Toxicity. In a developmental toxicity study, pregnant female rats were exposed by inhalation to brass flakes at 100 mg/m^3 for 15 or 150 min per day for 10 days on gestation days 6-15, and fetuses were collected and observed on gestation day 20 (Starke et al. 1987). Maternal body-weight gain was significantly reduced, as were the percentage of pregnancies, number of implantation sites, number of live fetuses, and body weights of fetuses in the group exposed for 150 min per day compared with the group exposed for 15 min per day or the control group. The data show unusual defects that the subcommittee believes can be classified as malformations (i.e., underdeveloped atria and thin-walled right cardiac ventricle). The malformations were suggestive of an effect of exposure. The disruption of implantation, in utero death, reduced maternal and fetal weight, and an increase in unusual defects indicated serious consequences of exposure to brass flakes for the longest exposure period. Thus, there is evidence of developmental toxicity in rats exposed to brass flakes at 100 mg/m^3 for 150 min per day during gestation days 6-15.

In a dominant lethal mutation study, male rats were exposed by inhalation to brass flakes at 100 mg/m^3, for 15 or 150 min per day, 5 days per week for 10 weeks (Starke et al. 1987). Twelve males per group were exposed and then mated to unexposed females. Only three males survived 150 min per day of exposure. The three males were pooled with the surviving males from a two-generation study (see below) that delayed the mating of the males in the dominant lethal mutation study for 4 weeks longer than that of the males exposed for 15 min per day or the controls. The significant increase in resorptions in pregnancies resulted from matings with males exposed for 150 min per day compared with those exposed for 15 min per day and the controls. A limitation of the study is that the number of offspring rather than the number of males should have been the group size used in the analysis of the data. That number might have resulted in a misinterpretation of the data. The severe general toxicity to male rats exposed at 100 mg/m^3 for 150 min per day, 5 days per week for 10 weeks compromises the interpretation of the data on reproductive or developmental effects in the study. There were no clear differences between the male rats exposed at 100 mg/m^3 for 15 min per day, 5 days per week for 10 weeks and the controls.

A third study, although called a two-generation study, was not a standard two-generation study because only the F_0 males and females were exposed to brass flakes (Starke et al. 1987). F_1 and F_2 generations were derived from the exposed parents but were not directly exposed. Twelve males and at least 24 females per group were exposed by inhalation to brass flakes at 100 mg/m^3 for 15 or 150 min per day, 5 days per week for 10 weeks (males) or 3 weeks (females) before mating. No significant difference in reproduction and body weight was observed in the groups exposed for 15 min per day and the control groups. Only 6 of the 12 males and 11 of the 24 females survived the exposure duration of 150 min per day. Because so few males and females survived, the observed effects on reproduction and body weight were difficult to interpret. Only 3 of 11 mated females gave birth, and it is not known whether the remaining females had been pregnant but had not given birth or whether they had never been pregnant. The effect of variability between mated males or litters was not addressed in this study, making it difficult to interpret the results. The authors concluded that there were no differences in viability and survival in the F_1 and F_2 generations; however, this conclusion is limited because the animals were only exposed in the F_0 generation. Although a nonstandard study design was used, the study does have relevance to the conditions of exposure in the military. Military personnel might be exposed to brass flakes; however, it is unlikely that their offspring will be similarly exposed.

Dermal, Ocular, and Oral Exposures

One-Time Exposures

Dermal and ocular studies of brass flakes are limited to two reports. Manthei et al. (1983) applied brass flakes suspended in corn oil to shaved adult rabbits at 2.0 g/kg. The application was allowed to remain on the skin for 24 hr. The brass-flake suspension caused mild erythema and edema, but no significant irritation. Muni et al. (1985) applied brass flakes (0.5 g) to gauze pads, moistened with saline, which were placed on two abraded and two unabraded rabbit skin sites and covered for 24 hr. The brass caused very slight edema to the unabraded site and slight edema to the abraded site; the edema resolved in both cases after 72 hr. Thus, EA 5763 appears to cause only minor skin irritation after a single dermal exposure.

Ocular toxicity was evaluated by Manthei et al. (1983) in rabbits by applying 100 mg of brass flakes in one eye and observing ocular irritation for 24, 48, and 72 hr and at 7 days after application. EA 5763 was classified as a positive eye irritant in these tests, but the damage was reversible. Muni et al. (1985) applied brass flakes (80 mg) to the right eye of three rabbits. Varying degrees of corneal opacity were produced in two animals within 24 hr. It resolved in 48 hr in one rabbit and 7 days in another rabbit. The brass was considered mildly irritating to the eye.

The toxicity of brass flakes by the oral route was determined by Manthei et al. (1983). EA 5763 was administered in corn oil to groups of 10 Sprague-Dawley rats (five of each gender) by gavage at doses of 5, 3.20, 2.00, 1.26, 0.79 and 0.50 g/kg. The high dose was lethal to four females and three males 48 to 144 hr after dosing. The lowest two doses were not lethal to any animals within 14 days of dosing. Toxicity was related to gastrointestinal distress and diarrhea. The 14-day lethal dose for 50% of the test animals (LD_{50}) was determined to be 2.78 g/kg (2.00-3.84 g/kg, 95% confidence limit).

An additional study by Muni et al. (1985) evaluated the oral 14-day LD_{50} of copper-zinc-coated powder in F344 albino rats. Range-finding studies using groups of two animals showed that oral (gavage) doses at 3,500 mg/kg were lethal to two of two males and two of two females, but doses at 1,050 mg/kg or less (315 or 94.5 mg/kg) were not lethal. When larger groups of 10 animals per group were evaluated, only two of five females and none of the males died within 14 days of administration of 3,300 mg/kg. The doses that were evaluated in the study with 10 animals per group were 3,300, 1,980, and 1,188 mg/kg, suspended in corn oil and administered by gavage. The doses lower than 3,300 mg/kg were not lethal to either gender. Thus, an approximate LD_{50} was determined to be greater than 3,300 mg/kg in this study.

Repeated Exposures

No studies are available concerning dermal, ocular, or oral toxicity of brass flakes after repeated exposures.

In Vitro Studies

No data are available concerning the mutagenicity of brass flakes in the Ames assay or other in vitro mutagenesis assays. A tracheal organ culture assay evaluated the cytotoxicity of brass dust to hamster upper respiratory epithelium (Placke and Fisher 1987). The brass particles caused severe degeneration and necrosis in the tracheal cultures at concentrations of 1,000 mg/L or higher.

SUMMARY OF TOXICITY DATA

Tables 3-1 to 3-3 summarize the brass-flakes toxicity studies. Table 3-1 presents some of the data on nonmammalian species and data on dermal, ocular, and oral exposures of rats. Those numbers will not be used to set guidance levels for inhalation exposures to brass. Table 3-2 presents data on single inhalation exposures to brass flakes in several species, and Table 3-3 presents data on multiple inhalation exposures to brass flakes in several species, including exposures of humans to brass dust.

Noncarcinogenic Effects

Acute exposure to high concentrations of brass flakes produces serious pulmonary inflammation and death in several species. The species most sensitive to the toxicity of brass flakes is the guinea pig (Thomson et al.

TABLE 3-1 Summary of Effects of a Single Oral, Ocular, or Dermal Exposure to Brass Flakes

Species	Exposure (route)	NOAEL	LOAEL	End Point and Comments	Reference
Rat	0.5-5.0 g/kg (oral)	—	0.5 g/kg	LD_{50} = 2,780 mg/kg oral dose of EA 5763; toxicity related to gastrointestinal distress	Manthei et al. 1983
Rabbit	100 mg in cornea (ocular)	—	100 mg	Reversible ocular irritation	Manthei et al. 1983
Rabbit	2.0 g/kg (dermal)	2.0 g/kg	—	Mild erythema, but not considered significant	Manthei et al. 1983
Rabbit	0.5 g to abraded or unabraded skin (dermal)	—	—	Very slight edema on un-abraded skin and slight edema on abraded skin; edema resolved after 72 hr	Muni et al. 1985
Rabbit	80 mg in cornea (ocular)	—	—	Corneal opacity produced in 2 out of 3 animals, resolved in 48 hr or 7 d	Muni et al. 1985
Rat	94.5-3,500 mg/kg (oral by gavage)	—	94.5 mg/kg	Body weight decreased at all doses; significant gastro-intestinal toxicity; oral LD_{50} > 3,300 mg/kg; material is Cu-Zn powder, not flakes	Muni et al. 1985

Abbreviations: NOAEL, no-observed-adverse-effect level; LOAEL, lowest-observed-adverse-effect level.

TABLE 3-2 Summary of Effects of a Single Inhalation Exposure to Brass Flakes

Species	Exposure	NOAEL (mg/m^3)	LOAEL (mg/m^3)	End Point and Comments	Reference
Rat	100 mg/m^3; 2.5 hr	—	100	BAL, LDH, ALKP, protein increased; inflammation increased	Thomson et al. 1982a
Guinea pig	100 mg/m^3; 150 min	—	100	100% lethality (12/12)	Thomson et al. 1982b
Rat	1,000 mg/m^3, 30, 50, 60, 70, 90 min	—	1,000 for 30 min (CT = 30,000 mg·min/m^3)	Lethality: at 50 min, 8/12 died; LCt$_{50}$ = 53,538 mg·min/m^3	Feeney et al. 1983
Mouse	1,000 mg/m^3; 15, 30, 120, 180, 300 min	—	1,000 for 30 min	Lethality: 1/12 at 1,000 mg/m^3, 30 min LCt$_{50}$ = 199,511 mg·min/m^3	Feeney et al. 1983
Rat	1, 10, 50, 100, 200 mg/m^3; 4 hr	1 (CT = 240 mg·min/m^3)	10 (CT = 2,400 mg·min/m^3)	10 mg/m^3 showed BAL changes in LDH and protein; no BAL changes at lowest dose; increase in respiratory resistance at 10 mg/m^3	Thomson et al. 1985; Thomson et al. 1986
Rat	100 mg/m^3; 4 hr	—	100	BAL showed significant inflammation	Anderson et al. 1988
Rat	100 mg/m^3; 4 hr	—	100	Respiratory-tract lesions (no data presented)	Yeh et al. 1990

Abbreviations: NOAEL, no-observed-adverse-effect level; LOAEL, lowest-observed-adverse-effect level; BAL, bronchoalveolar lavage; LDH, lactate dehydrogenase; ALKP, alkaline phosphatase; LCt$_{50}$, lethal concentration for 50% of the test animals multiplied by exposure time; CT, product of exposure concentration and time.

TABLE 3-3 Summary of Effects of Multiple Inhalation Exposures to Brass Flakes

Species	Exposure	NOAEL (mg/m^3)	LOAEL (mg/m^3)	End Point and Comments	Reference
Human	Occupational, combined Cu + Zn; at 0.001-0.191 mg/m^3; 10-17 yr	0.001	0.19	Chronic bronchitis, asthma, related to length of exposure >10 yr; some dose response evident	Rastogi et al. 1991, 1992a,b
Rat	100 mg/m^3; 2.5 hr, 6 d	—	100	BAL used, multiple exposures much more toxic than single exposure	Thomson et al. 1982a
Rat	100 mg/m^3; 150 min/d, 5 d/wk, 6 wk	—	100	Lethality: 23/72 after 7 d; 25-30% decreased body-weight gain; exposures terminated after 7 d	Thomson et al. 1982b
Rat	100 mg/m^3; 150 min/d, 5 d/wk, 13 wk	—	100	Exposures terminated after 7 d because of lethality	Thomson et al. 1982b
Rat	100 mg/m^3; 15 min/d, 5 d/wk, 6 wk	—	100	Marked pulmonary lesions in 4/12	Thomson et al. 1982b
Rat	100 mg/m^3; 15 min/d, 5 d/wk, 13 wk	—	100	Lethality: 9/72, 2nd - 9th w; decreased tidal volume and lung compliances (continued in males 30 d after exposures); mild pulmonary lesions in all 30 d after exposure	Thomson et al. 1982b
Mouse	100 mg/m^3; 150 min/d, 5 d/wk, 6 wk	—	100	Not lethal, dyspnea, hypo-activity, lower body weight	Thomson et al. 1982b
Mouse	100 mg/m^3; 150 min/d, 5 d/wk, 13 wk	—	100	Respiratory lesions	Thomson et al. 1982b

TABLE 3-3 *(Continued)*

Species	Exposure	NOAEL (mg/m³)	LOAEL (mg/m³)	End Point and Comments	Reference
Mouse	100 mg/m³; 15 min/d, 5 d/wk, 6 wk	100	—	No significant pulmonary lesions	Thomson et al. 1982b
Mouse	100 mg/m³; 15 min/d, 5 d/wk, 13 wk	—	100	Slight damage after 13 wk	Thomson et al. 1982b
Guinea pig	100 mg/m³; 150 min/d, 5 d/wk, 13 wk	—	100	Lethality: 12/12 after 150 min of exposure	Thomson et al. 1982b
Rat	1 and 10 mg/m³; 6 h/d, 5 d/wk, 6 wk and 13 wk	—	1	High dose: increases respiration rate, significant lesions; low dose: mild, reversible lesions	Thomson et al. 1984
Rat	0.32, 1.0, 3.2, 10 mg/m³; 1.5 h/d, 4 d/wk, 13 wk	0.32	1.0	1.0 mg/m³ mild focal atrophy olfactory epithelium (resolved in 4 wk)	Snipes et al. 1988
Rat (males)	100 mg/m³; 15 or 150 min/d, 5 d/wk, 10 wk	—	100	At 15 min/d, no reproductive or developmental effects observed in offspring of exposed males; at 150 min/d most exposed animals died during exposure, significant increase in resorption of pregnancies; results difficult to interpret	Starke et al. 1987
Rat (pregnant females)	100 mg/m³; 15 or 150 min/d, gestation days 6-15	—	100	Evidence of developmental toxicity at 100 mg/m³, 150 min/d	Starke et al. 1987

Rat (males and females)	F_0 generation exposed at 100 mg/m³; 15 or 150 min/d, 5 d/wk, 10 wk (males) or 3 wk (females)	100	—	No differences on viability and survival in F_1 and F_2 generations; results difficult to interpret because only rats in F_0 generation exposed and few survived the exposure period	Starke et al. 1987

Abbreviations: NOAEL, no-observed-adverse-effect level; LOAEL, lowest-observed-adverse-effect level; BAL, bronchoalveolar lavage.

1982b). However, few studies have been conducted with this species. The only available data on this species were from exposures to a single concentration (100 mg/m^3) for a single duration (150 min) that was lethal to all 12 animals. Thus, results concerning the lowest dose that produces toxicity are too uncertain to permit the use of data from guinea pigs to set guidance levels. Rats are more sensitive than mice to the toxic effects of brass flakes (Feeney et al. 1983), and respiratory effects are the predominant toxic end points in this species. One-time exposure to brass flakes is lethal to rats at LCt$_{50}$ values of approximately 50,000 mg•min/m^3 (Feeney et al. 1983). Functional respiratory deficits and enzymatic and cytological changes in bronchopulmonary lavage fluid were observed in rats exposed at 2,400 mg•min/m^3 (Thomson et al. 1986). Exposure at 1 mg/m^3 for 4 hr (CT, 240 mg•min/m^3) was a NOAEL for rats, and that value can be used to calculate guidance levels for a one-time exposure to brass flakes.

Multiple exposures to brass flakes significantly increased the toxicity of this material to rats and mice (Thomson et al. 1982b). Exposure of rats at 100 mg/m^3 for 4 hr (CT, 24,000 mg•min/m^3) produced no lethality, but the same concentration (100 mg/m^3) for only 15 min per day, 5 days per week (cumulative CT, 15,000 mg•min/m^3 by the end of week 2 was lethal to 9 of 72 rats during the second through the ninth week of a chronic exposure (Thomson et al. 1982b). Longer daily exposures for 150 min per day, 5 days per week (cumulative CT, 75,000 mg•min/m^3 by the end of week 1) were lethal to 23 of 72 rats. A careful study (Snipes et al. 1988) with many biochemical, functional, and histological end points demonstrated a NOAEL in rats of 0.32 mg/m^3 when exposures were conducted for 90 min per day, 4 days per week for 13 weeks (cumulative CT, 1,500 mg•min/m^3 by the end of week 13). That value can be used to calculate guidance levels for multiple exposures to brass flakes.

Carcinogenic Effects

No studies have been done on the carcinogenicity of brass flakes.

PREVIOUS RECOMMENDED EXPOSURE LIMITS

There are no current recommended exposure limits for brass flakes as used by the military in obscurant grenades. However, brass flakes are composed of approximately 70% Cu and 30% Zn, so some comparisons can be made to current standards or guidelines for exposure to Cu and Zn from various agencies. The current Occupational Safety and Health Administration (OSHA) permissible exposure limits (PELs) for Zn oxide are 5 mg/m^3 for fumes and 10 mg/m^3 for dusts. The OSHA PELs for Cu are 0.1 mg/m^3 for fumes and 1.0 mg/m^3 for dusts. The Threshold-Limit-Value–time-weighted-average (TLV-TWA) values established for Cu by the American Conference of Government Industrial Hygienists (ACGIH) are 0.2 mg/m^3 for fumes and 1.0 mg/m^3 for dust.

The brass flakes evaluated in this chapter are not considered fumes because they are approximately 2 μm in diameter and because the oxides of fumes are chemically different from brass flakes. Cohen and Powers (1994) reported that the particle size of dusts in a nonferrous brass-casting foundry was greater than the particle size in fumes. Therefore, an appropriate comparison between the brass flakes used by the military and the OSHA PELs or ACGIH TLV-TWAs for Cu and Zn dusts cannot be made.

SUBCOMMITTEE EVALUATION AND RECOMMENDATIONS

Using the toxicity information described above, the subcommittee recommended exposure guidance levels for military personnel exposed during an emergency release and during regular training exercises and for consideration at military-facility boundaries to protect nearby communities from emergency or repeated releases of brass flakes.

MILITARY EXPOSURES

Emergency Exposure Guidance Levels (EEGLs)

Acute inhalation exposure to high concentrations of brass flakes is lethal to all animal species that have been tested. Respiratory effects are the first to appear. Pulmonary inflammation can be produced in rats with 4-hr exposures at a concentration of 10 mg/m^3 (Thomson et al. 1986); however, that response is reversible, and clearance of the particulate matter appears to be fast (Muse 1983).

The 4-hr NOAEL of 1 mg/m^3 identified by Thomson et al. (1986) for

F344 rats is used to recommend EEGLs for 15 min, 1 hr, and 6 hr. An uncertainty factor of 10 is used to extrapolate from animals to humans. Although no studies have demonstrated that Haber's rule can be applied to brass flakes, the mechanism of acute toxicity appears to be related to inflammatory responses. Acute inflammation of airways should follow Haber's rule for this material. Therefore, an increase in acceptable concentrations for shorter durations is warranted. Thus the 1-hr and 6-hr EEGLs for brass flakes are 0.4 mg/m^3 and 0.07 mg/m^3, respectively. Extrapolating back from the 1-hr EEGL indicates a 15-min EEGL of 1.6 mg/m^3. However, on the basis of Thomson et al. (1986) data, extrapolating back from a 4-hr NOAEL to a 15-min exposure might exceed the range of exposure durations over which Haber's rule can be expected to apply. The study by Feeney et al. (1983) indicated that exposure at 1,000 mg/m^3 for 30 min was not lethal to rats but did produce a variety of pathological lesions of moderate severity in the respiratory tract. That value is used to check the validity of the proposed 15-min EEGL. A LOAEL-to-NOAEL uncertainty factor of 10, a species uncertainty factor of 10, and an uncertainty factor of 10 to extrapolate from moderate toxic effects to mild reversible effects are applied to estimate a 30-min exposure guidance level of 1 mg/m^3, which agrees well with the recommended 15-min EEGL of 1.6 mg/m^3, assuming Haber's rule applies over the 15-min time interval. Thus, the recommendation appears to be validated by the 30-min exposure study.

Repeated Exposure Guidance Level (REGL)

Multiple exposures to brass flakes produce significantly more toxicity in experimental animals than single exposures that have the same cumulative exposure dose. Data are not available on the effects of human exposure to brass flakes; however, humans who were occupationally exposed to high concentrations of brass fumes exhibit an increased risk of chronic bronchitis and other severe respiratory problems. Combined Cu and Zn concentrations that were shown to be associated with chronic bronchitis were 0.19 mg/m^3 over a period of at least 10 years. Concentrations that could not be associated with respiratory disease were about 0.001 mg/m^3 (Rastogi et al. 1991).

The human NOAEL is 0.001 mg/m^3 (Rastogi et al. 1991), and the subcommittee recommends that value as the REGL. The REGL is supported by two experimental animal studies. Snipes et al. (1988) exposed rats to brass flakes at concentrations of 0.32, 1.0, 3.2, or 10 mg/m^3 for 1.5 hr per day, 4 days per week for 13 weeks and identified a NOAEL of 0.32 mg/m^3. Applying a time extrapolation of 6 hr/40 hr to that concentration, a species uncertainty factor of 10, and a 10-fold uncertainty factor to extrapolate from a 13-week subchronic exposure to a chronic exposure, a REGL of 0.0005 mg/m^3 is obtained. That value is lower than the recommended REGL of 0.001 mg/m^3 by only a factor of 2. Thomson et al. (1984) exposed rats to brass flakes at 1.0 or 10 mg/m^3 for 6 hr per day, 5 days per week for 13 weeks and identified a LOAEL of 1.0 mg/m^3. Applying a time extrapolation of 30 hr/40 hr for that concentration, a species uncertainty factor of 10, a 10-fold uncertainty factor to extrapolate from a 13-week subchronic exposure to a chronic exposure, and a 10-fold uncertainty factor to extrapolate from a LOAEL to a NOAEL, a REGL of 0.0008 is obtained. That value is lower than the recommended REGL by only a factor of 1.25. Therefore, the subcommittee concludes that the data from the two experimental animal studies provide justification for the use of the human NOAEL value in the derivation of the REGL.

PUBLIC EXPOSURES

Short-Term Public Emergency Guidance Levels (SPEGLs)

The subcommittee assumes that more inherent susceptibility to the respiratory effects of brass flakes will be found in the general population than in healthy adult military personnel. Thus, an additional uncertainly factor of 10 is used to extrapolate from the recommended EEGLs to levels that protect the general public. Therefore, the SPEGLs for a single emergency exposure to brass flakes are 0.16, 0.04, and 0.007 mg/m^3 for 15 min, 1 hr, and 6 hr, respectively.

Repeated Public Exposure Guidance Level (RPEGL)

A RPEGL is estimated from the REGL by incorporating an additional uncertainty factor of 10 to protect susceptible individuals in the general population. Thus, the RPEGL is 0.0001 mg/m^3.

SUMMARY OF SUBCOMMITTEE RECOMMENDATIONS

The subcommittee's recommendations for exposure limits for military personnel exposed to brass flakes are listed in Table 3-4. Table 3-5 summarizes the subcommittee's recommendations for exposure limits for the general public. The subcommittee recommends that the Army's use of brass flakes be limited to minimize the potential for increasing toxicity with repeated exposures.

RESEARCH NEEDS

The subcommittee recommends that long-term, chronic inhalation studies in experimental animals be conducted to characterize the precise toxic effects due to chronic exposure. Additionally, reproductive and developmental studies using standard testing protocols should be considered.

TABLE 3-4 EEGLs and REGL for Brass Smoke for Military Personnel

Exposure Guideline	Exposure Duration	Exposure Guidance Level (mg/m^3)
EEGL	15 min	1.6
	1 hr	0.4
	6 hr	0.07
REGL	8 hr/d, 5 d/wk	0.001

Abbreviations: EEGL, emergency exposure guidance levels; REGL, repeated exposure guidance level.

TABLE 3-5 SPEGLs and RPEGL for Brass Smoke for the Boundaries at Military-Training Facilities

Exposure Guideline	Exposure Duration	Exposure Guidance Level (mg/m^3)
SPEGL	15 min	0.16
	1 hr	0.04
	6 hr	0.007
RPEGL	8 hr/d, 5 d/wk	0.0001

Abbreviations: SPEGL, short-term public emergency guidance level; RPEGL, repeated public exposure guidance levels.

REFERENCES

Anderson, R.S., L.L. Gutshall, and S.A. Thomson. 1988. Responses of rat alveolar macrophages to inhaled brass powder. J. Appl. Toxicol. 8:389-393.

Cohen, H.J., and B.J. Powers. 1994. A study of respirable versus nonrespirable copper and zinc oxide exposures at a nonferrous foundry. Am. Ind. Hyg. Assoc. J. 55:1047-1050.

Feeney, J.J., P. Hott, R.L. FarranD, and J.T. Weimer. 1983. Acute Inhalation Toxicity of EA 5769 in Rats and Mice Comparison of Toxicity with EA 5763. ARCSL-TR-82069. Chemical Systems Laboratory, U.S. Army Armament, Munitions and Chemical Command, Aberdeen Proving Ground, Edgewood, Md.

Grose, E.C., and J.A. Graham. 1987. Short-term In Vitro Screening Studies Related to the Inhalation Toxicology of Potentially Toxic Aerosols. Procurement No. 84PP4850. Prepared by U.S. Environmental Protection Agency, Health Effects Research Laboratory, Research Triangle Park, N.C., for U.S. Army Medical Bioengineering Research and Development Laboratory, Fort Detrick, Md.

Haley, M.V., and C.W. Kurnas. 1993. Toxicity and Fate Comparison Between Several Brass and Titanium Dioxide Powders. ERDEC-TR-094. Edgewood Research, Development and Engineering Center, U.S. Army Armament, Munitions and Chemical Command, Aberdeen Proving Ground, Edgewood, Md.

Manthei, J.H., F.K. Lee, Jr., M. Donnelly, and J.T. Weimer. 1980. Preliminary Toxicity Screening Studies of 11 Smoke Candidate Compounds. ARCSL-TR-79056. Chemical Systems Laboratory, U.S. Army Armament, Munitions and Chemical Command, Aberdeen Proving Ground, Edgewood, Md.

Manthei, J.H., F.K. Lee, D.H. Heitkamp, and W.C. Heyl. 1983. Preliminary and Acute Toxicological Evaluation of Five Candidate Smoke Compounds. ARCSL-TR-82066. Chemical Systems Laboratory, U.S. Army Armament, Munitions and Chemical Command, Aberdeen Proving Ground, Edgewood, Md.

Muni, I.A., E.B. Gordon, and J.B. Goodband. 1985. Dermal, Eye, and Oral Toxicological Evaluations. Phase II Report. DAMD17-82-C-2301. Prepared by Bioassay Systems Corp., Woburn, Mass., for U.S. Army Research and Development Command, Fort Detrick, Frederick, Md.

Muse, W.T. 1983. A Pilot Study to Determine the Biological Fate of Inhaled EA 5769 in Rats. ARCSL-TR-82067. Chemical Systems Laboratory, U.S. Army Armament, Munitions and Chemical Command, Aberdeen Proving Ground, Edgewood, Md.

Placke, M.E., and G.L. Fisher. 1987. In Vitro Toxicity Evaluation of Ten Particulate Materials in Tracheal Organ Culture. CRDEC-TR-88010. Chemical Research, Development and Engineering Center, U.S. Army Armament, Munitions and Chemical Command, Aberdeen Proving Ground, Edgewood, Md.

Rastogi, S.K., B.N. Gupta, T.H. Husain, B.S. Pangtey, S. Srivastava, and N. Garg.

1991. Long-term effects of soldering fumes upon respiratory symptoms and pulmonary function. Am. Occup. Hyg. 35:299-307.

Rastogi, S.K., B.N. Gupta, N. Mathur, T. Husain, P.N. Mahendra, B.S. Pangtey, and S. Srivastava. 1992a. A survey of chronic bronchitis among brassware workers. Am. Occup. Hyg. 36:283-294.

Rastogi, S.K., B.N. Gupta, T. Husain, N. Mathur, B.S. Pangtey, and N. Garg. 1992b. Respiratory symptoms and ventilatory capacity in metal polishers. Hum. Exp. Toxicol. 11:466-472.

Snipes, M.B., D.E. Bice, D.G. Burt, E.G. Damon, A.F. Eidson, F.F. Hahn, B.R. Harkema, R.F. Henderson, J.L. Mauderly, J.A. Pickrell, F.A. Seiler, and H.C. Yeh. 1988. Comparative Inhalation Toxicology of Selected Materials—Phase III Studies. Procurement No. 85PP5805. Prepared by Inhalation Toxicology Research Institute, Lovelace Biological and Environmental Research Institute, Albuquerque, N.M., for U.S. Army Biomedical Research and Development Laboratory, Fort Detrick, Frederick, Md.

Starke, W.C., R.J. Pellerin, D.C. Burnett, C.J. Richmond, J.H. Manthei, and D.H. Heitkamp. 1987. Inhalation of Brass Flakes, Effects on Reproduction in Rats. CRDEC-TR-88035. Chemical Research, Development and Engineering Center, U.S. Army Armament, Munitions and Chemical Command, Aberdeen Proving Ground, Edgewood, Md.

Thomson, S., J. Height, F. Lee, A. Cooper, and L. Buettner. 1982a. Enzymatic and Cytological Changes in Lung Lavage Fluid from Rats Exposed to Inhaled EA 5763 Flakes. ARCSL-TR-81049. Chemical Systems Laboratory, U.S. Army Armament, Munitions and Chemical Command, Aberdeen Proving Ground, Edgewood, Md.

Thomson, S.M., J. Callahan, C. Crouse, D. Burnett, R. Pellerin, J. Height, R. Farrand, and D. Heitkamp. 1982b. The Effects of Subchronic Toxicity in Rats and Mice Exposed by Inhalation to EA 5763 (100 mg/cu m). ARCSL-TR-82026. Chemical Systems Laboratory, U.S. Army Armament, Munitions and Chemical Command, Aberdeen Proving Ground, Edgewood, Md.

Thomson, S.M., D.C. Burnett, J.F. Callahan, C. Crouse, A. Cooper, J.J. Height, D.H. Heitkamp, R.L. Farrand, W. Muse, and R.J. Pellerin. 1984. Subchronic Inhalation Toxicity of EA 5763 (1 and 10 mg/m^3) in Rats. CRDC-TR-84027. Chemical Research, Development and Engineering Center, U.S. Army Armament, Munitions and Chemical Command, Aberdeen Proving Ground, Edgewood, Md.

Thomson, S.M., D. Burnett, J. Bergmann, M. Lamb, J. James, A. Cooper, and L. Yellets. 1985. A study of the Acute Inhalation Hazards of EA 5752, EA 5763, and EA 5763 (Dedusted) Using Bronchopulmonary Lavage in the Rat. CRDC-TR-84122. Chemical Research, Development and Engineering Center, U.S. Army Armament, Munitions and Chemical Command, Aberdeen Proving Ground, Edgewood, Md.

Thomson, S.M., D.C. Burnett, J.D. Bergmann, and C.J. Hixon. 1986. Comparative inhalation hazards of aluminum and brass powders using bronchopulmonary lavage as an indicator of lung damage. J. Appl. Toxicol. 6:197-209.

Wentsel, R.S. 1986. Fate and Effects of Brass Powder on the Environment. CRDEC-TR-86044. Chemical Research, Development and Engineering Center, U.S. Army Armament, Munitions and Chemical Command, Aberdeen Proving Ground, Edgewood, Md.

Yeh, H.C., M.B. Snipes, A.F. Eidson, C.H. Hobbs, and M.C. Henry. 1990. Comparative evaluation of nose-only versus whole-body inhalation exposures for rats—Aerosol characteristics and lung deposition. Inhalation Toxicol. 2:205-221.

4
Titanium Dioxide Smoke

BACKGROUND INFORMATION

MILITARY APPLICATIONS

TITANIUM DIOXIDE (TiO_2) is the proposed major component of a training grenade—XM82—under development by the U.S. Army Chemical Research Development and Engineering Center. TiO_2 particles will be used to block detection of light waves in the visible portion of the electromagnetic spectrum. In using this training grenade, military personnel are likely to be exposed to airborne TiO_2 particles.

PHYSICAL AND CHEMICAL PROPERTIES

TiO_2 is a noncombustible, white crystalline solid. Common crystalline forms are anatase and rutile, the latter being the more thermally stable. Commercially available TiO_2 products are being considered for use by the U.S. Army. The commercial products can be composed of the pure anatase or rutile forms or a mixture of both forms. Both forms of TiO_2 can also occur in nature. TiO_2 has a molecular weight of 79.90 and specific gravity of 3.90 (anatase) and 4.23 (rutile). The melting point for TiO_2 is 1,830-1,850°C, and the boiling point is 2,500-3,000°C. TiO_2 is insoluble in water, hydrochloric acid, nitric acid, or alcohol. It is soluble in hot concentrated sulfuric acid, hydrogen fluoride, or alkali. TiO_2 can exist as a fine- or ultrafine-sized particle.[1] Fine particles are defined as material with individual particle diameters of 0.1-2.5 micrometer (μm). Ultrafine particles are defined as particles with individual diameters of <0.1 μm. Both fine and ultrafine TiO_2 particles can exist as aggregates of the individual particles.

OCCURRENCE AND USE

TiO_2 (anatase and rutile forms) is widely used as a pigment in paints, varnishes, and lacquers and as a pigment and filler for paper. It is used as an additive in polymer production and electronic-component production and as a catalyst. The anatase and rutile forms are used as a welding-rod coating and a ceramic colorant. TiO_2 is used in the formulation of topical and oral pharmaceuticals and in food colorants.

TOXICOKINETICS

The clearance and distribution of TiO_2 particles in rats have been examined after acute and chronic inhalation and after acute intratracheal instillation. Ferin and Oberdörster (1985) characterized lung clearance of the two most common forms of TiO_2, anatase and rutile, in rats after a 7-hr exposure to anatase at 16.5 milligrams per cubic meter (mg/m^3) or rutile at 19.3 mg/m^3. The materials were fine-sized particles. Lung clearance half-times for anatase and rutile were 51 and 53 days, respectively. The results indicated no difference in the lung clearance of both forms of TiO_2.

Rats were exposed by inhalation 6 hr per day, 5 days per week for 12 weeks to ultrafine-sized (diameter 0.02 μm) and fine-sized (diameter 0.25 μm) TiO_2 particles at approximately 23 mg/m^3 (Ferin et al. 1992). At the end of the 12-week exposure, the total lung-particle burdens were

---

[1]Three sizes of airborne particles are frequently described, especially for ambient particulate matter. The sizes are ultrafine (<0.1 μm), fine (0.1-2.5 μm), and coarse (>2.5 μm). For ambient particulate matter, the sizes reflect not only differences in size but also differences in formation and composition.

similar for the fine and ultrafine TiO_2 preparations. However, more ultrafine particles were found in the lung-associated lymph nodes and the lung interstitium. Lung-particle clearance was also significantly slower for the ultrafine TiO_2 particles than for the fine; clearance half-times were 501 days and 174 days, respectively. Similar to observations made after inhalation exposure, intratracheal instillation of ultrafine and fine TiO_2 particles was associated with greater interstitialization of the ultrafine particles (Ferin et al. 1992). These inhalation and intratracheal instillation studies demonstrated that the size of the TiO_2 particles, particularly, ultrafine particles can influence distribution and clearance behavior in the rat lung.

The distribution of fine-particle (diameter >0.1 μm) TiO_2 in rats exposed for 24 months by inhalation to 10, 50, and 250 mg/m^3 was described by Lee et al. (1985a). TiO_2 was found throughout the respiratory tract, although the greatest accumulation was in macrophages localized at the terminal bronchioles and alveolar ducts and in lung-associated lymph nodes. Some TiO_2 particles were visualized in the interstitium and in alveolar epithelial cells. TiO_2 was found in the liver; the amount of material present was related to exposure concentration. The greatest amount of particles was in the peripheral hepatic lobules. The spleen also showed exposure-related dust deposition, primarily in the lymphoid tissue of the white pulp. The results of this study showed that the particles can be found throughout the respiratory tract and in extra-respiratory tissues after chronic exposure to high concentrations of TiO_2.

TOXICITY SUMMARY

EFFECTS IN HUMANS

Chen and Fayerweather (1988) conducted an analysis of lung-cancer mortality and incidence of nonmalignant respiratory disease in a cohort of 1,576 workers exposed to TiO_2 for at least 1 year in two production facilities between 1935 and 1984. No significant association was observed between TiO_2 exposure and malignant or nonmalignant respiratory disease. Time-weighted-average (TWA) exposure concentrations of TiO_2 in the plants ranged from 1 to 20 mg/m^3.

Effects in Animals

Inhalation Exposures

One-Time Exposures

Lethality. The lethal inhaled concentration of TiO_2 in rats is reported to be >6820 mg/m^3 for a 4-hr exposure (DuPont 1979). Rats exposed for 4 hr at 6820 mg/m^3 exhibited irregular respiration, gasping, and lethargy; however, no deaths occurred during a 14-day observation period after exposure (DuPont 1979).

Pulmonary Effects. See Table 4-1 for a summary of the nonlethal pulmonary effects of exposure to TiO_2 by inhalation.

Hilaski et al. (1992) exposed rats for 30 min to aerosols of fine TiO_2 particles generated with a XM82 grenade. Responses were characterized by histopathological analysis and bronchoalveolar-lavage (BAL) fluid analyses and pulmonary-function tests. Two sets of high-concentration exposures of rats were performed; one set was evaluated 24 hr after exposure and the other was evaluated 14 days after exposure. In the 24-hr evaluation, the TiO_2 concentrations determined 5, 15, and 25 min after detonating the grenade were 2,260, 1,506, and 1,000 mg/m^3. In the 14-day evaluation, exposure concentrations determined at 5, 15, and 25 min after detonating the grenade were 1,960, 1,300, and 964 mg/m^3, respectively. The TWA concentration for the 30-min exposure in the 14-day evaluation was 1,240 mg/m^3. The only consistent exposure-related response observed after 24 hr or 14 days was the presence of pigment-containing macrophages in the lung. Gases measured in the exposure chamber in the two sets of exposures included carbon monoxide (CO), carbon dioxide (CO_2), nitrogen oxide (NO), formaldehyde (HCHO), and trichloroethane (CH_3CCl_3). The concentrations of all of them, with the exception of CO and HCHO, were below the current occupational exposure limits. CO was measured slightly above the 50 parts per million (ppm) Threshold Limit Value (TLV) TWA and the HCHO concentrations were 3 and 2.5 ppm (TLV ceiling (C) = 0.3 ppm).

Ferin and Oberdörster (1985) compared the lung toxicity of anatase and rutile. Groups of rats were given a single intratracheal instillation exposure to saline or a saline suspension of 0.5 or 5 mg of anatase or rutile. Both forms of TiO_2 elicited an inflammatory response in the lung assessed 24 hr after exposure by BAL-fluid analysis. No remarkable

TABLE 4-1 Summary of Nonlethal Effects of Inhalation Exposure to Titanium Dioxide Smoke

Species	Exposure Frequency and Duration	NOAEL (mg/m³)	LOAEL (mg/m³)	End Point and Comments	Reference
Rat (male F344)	Single 30 min	1,240 (30 min, TWA)	—	TiO_2 generated from a XM82 grenade; particle size, MMAD 2.0-2.2 μm; no adverse effects observed at 24 hr or 14 d after exposure	Hilaski et al. 1992
Rat (female F344/N)	Single 4 hr	—	—	Particle size, MMAD 0.92-1.19 μm; lung burden was 217-252 μg/g lung; reported TiO_2 relatively nontoxic; however, no data shown	Yeh et al. 1990
Rat (male Long-Evans)	Single 7-hr inhalation; single i.t.	Inhalation: 16.5 anatase, 19.3 rutile; i.t., 0.5 anatase or rutile	—	Particle size, anatase MMAD 1.0 μm; rutile MMAD 0.83 μm; lung clearance half-times 51 and 53 d for anatase and rutile, respectively; no remarkable differences observed in anatase and rutile	Ferin and Oberdorster 1985
Rat (male F344)	4 hr/d, 4 d; 14-d followup	—	101.5	Particle size, MMAD 1.4-1.6 μm; transient increase in BAL fluid neutrophils; no adverse effects on pulmonary function detected; response to TiO_2 less than that to graphite	Thomson et al. 1988, 1990
Rat (male F344)	6 hr/d, 5 d	—	51.1	Particle size, MMAD 1.0 μm; no adverse effects detected	Driscoll et al. 1991
Rat (female F344)	6 hr/d, 5 d/wk, 4 wk inhalation; single i.t.; 24-wk followup	Inhalation, 10; i.t., 0.75	—	Particle size, MMAD 1.2-1.3 μm; fine particle size; no significant lung effects detected	Henderson et al. 1995

Species	Exposure schedule		Concentration (mg/m³)	Effects	Reference
Rat (male Crl:CDBR)	6 hr/d, 5 d/wk, 4 wk	5	250	Fine-sized TiO_2; inflammation and increased BrdU cell labeling observed after 250 mg/m³ but not 5 mg/m³	Warheit et al. 1996
Rat	Not described	—	—	Smoke from a TiO_2/hexachloroethane mixture much less toxic than smoke from a Zn/hexachloroethane mixture	Karlsson et al. 1985
Rat (F344)	7 hr/d, 5 d/wk, 1, 2, 4, 12 wk; 12-mo followup	—	10	Minimal pleural fibrosis observed after 12 wk of inhalation exposure	Johnson and Wagner 1989
Rat (male F344)	6 hr/d, 5d/wk, up to 12 wk, inhalation; single 65-1,000 µg, i.t.	—	Inhalation: 23.5 ultrafine, 23 fine	Ultrafine (i.e., <0.1 µm) and fine (1 µm) particles tested; greater retention and toxicity with ultrafine	Ferin et al. 1992; Oberdorster et al. 1994a
Rat (male F344)	6 hr/d, 5 d/wk, 12 wk	—	23.5 ultrafine, 22.3 fine	Ultrafine (~0.20 µm) and fine (0.25 µm) particles tested; ultrafine produced greater inflammation, fibrosis, and impairment of particle clearance	Oberdorster et al. 1994b
Rats (male and female CD)	6 hr/d, 5 d/wk, 24 mo	—	10	Fine particles tested; MMAD 1.5-1.7 µm; lung burden after 24 mo exposure 665 mg/lung at 250 mg/m³; some particles in extra-respiratory tissues; adverse effects confined to the respiratory tract and associated lymph nodes; increased lung tumors at highest exposure	Lee et al. 1985a,b; 1986

TABLE 4-1 *(Continued)*

Species	Exposure Frequency and Duration	NOAEL (mg/m^3)	LOAEL (mg/m^3)	End Point and Comments	Reference
Rat (F344)	6 hr/d, 5 d/wk, 24 mo	5	—	Fine particles tested; MMAD 1.1 μm; lung burden after 24 mo exposure 2.7 mg/lung; no adverse effects were detected by histopathology or analysis of BAL fluid	Muhle et al. 1991
Rat (female Wistar)	18 hr/d, 5 d/wk, 24 mo	—	10.4	Ultrafine particles (14 nm); lung burden 39.3 mg/lung; TiO$_2$ produced epithelial hyperplasia, fibrosis, and lung tumors; similar responses in rats exposed to carbon black	Heinrich et al. 1995
Rat (female Wistar)	18 hr/d, 5 d/wk, 24 mo	—	10.4	Ultrafine particles (14 nm) tested; impaired alveolar clearance of tracer particles detected after TiO$_2$, diesel soot and carbon black	Creutzenberg et al. 1990
Rat (female Wistar)	18 hr/d, 5 d/wk, 24 mo	—	10.4	Ultrafine particles (14 nm) tested; significant decrease in DNA adducts in lungs of rats exposed to TiO$_2$ at average of 10 mg/m^3	Gallagher et al. 1994
Rat (F344)	Single i.t.; 28-d followup	—	10 mg/kg body wt	Ultrafine and fine particles tested; ultrafine TiO$_2$ produced greater inflammatory response than the fine material	Driscoll and Maurer 1991
Rat (male F344)	Single i.t.; 28-d followup	—	5 mg/kg body wt	Fine particles tested; transient inflammatory response at 5 mg/kg body wt; fibrosis at 50 and 100 mg/kg; activation of macrophage cytokine release at 10, 50, and 100 mg/kg	Driscoll et al. 1990a,b

Species	Exposure	Dose	Effects	Reference
Rat (Sprague-Dawley)	Single i.t.; 2-d followup	—	i.t. TiO$_2$ resulted in transient inflammation and increases in BAL fluid desmosine	Li et al. 1996
Rat (F344)	10 and 100 mg/kg body wt, i.t.	10 mg/kg	TiO$_2$ at 100 mg/kg resulted in lung inflammation and increased mutation in alveolar type II cells	Driscoll et al. 1997
Rat (F344)	6 hr/d, 5 d/wk, 24 mo	100 mg/kg	Particle size, MMAD 1.1 μm; lung burden after 24 mo 2.7 mg/lung; no effect on the alveolar clearance of low solubility tracer particles	Bellmann et al. 1991
Mouse (female NMRI)	18 hr/d, 5 d/wk, 13.5 mo	5	Lung burden after 12 mo exposure 5.2 mg/lung; no increase in lung tumors after TiO$_2$, diesel soot, or carbon black	Heinrich et al. 1995
Mouse (CBA/ca)	20 hr/d, 2 or 4 wk	10.4	TiO$_2$ at 20 mg/m^3 for 2 or 4 wk impaired lung clearance of bacteria *Pasteurella haemolytica*; TiO$_2$ at 20 mg/m^3 appeared to suppress local immune responses to antigen	Gilmour et al. 1989a,b
Dog	i.t., frequency not given; 9-15 mo observation	20	TiO$_2$ administered by i.t.; dose was not given; alveolitis, centrilobular emphysema, focal collapse of alveoli, and fibroblast hyperplasia reported	Zeng et al. 1989
Human	Occupational	Estimated up to 20	Epidemiological study observed no increase in cancer, nonmalignant respiratory disease, or other diseases in TiO$_2$ exposed workers	Chen and Fayerweather 1988

Abbreviations: NOAEL, no-observed-adverse-effect level; LOAEL, lowest-observed-adverse-effect level; MMAD, mass median aerodynamic diameter; TWA, time-weighted average; i.t., intratracheal instillation; BAL, bronchoalveolar lavage; BrdU, 5-bromo-2'-deoxyuridine.

differences were found in the inflammatory activity of the two forms of TiO_2.

Driscoll et al. (1990a,b) exposed Fischer 344 (F344) rats by intratracheal instillation to TiO_2 at concentrations of 5, 10, 50, and 100 milligrams per kilogram (mg/kg) of body weight (body weights ranged from 180 to 220 grams (g), resulting in burdens of approximately 1, 2, 10, and 20 mg) and characterized lung responses by BAL-fluid and histopathological analyses 1, 7, 14, and 28 days after exposure. Exposure at 50 and 100 mg/kg (lung burdens of 10 and 20 mg, respectively) resulted in lung inflammation and fibrosis, exposure at 10 mg/kg (lung burden of 2 mg) produced minimal transient inflammation, and exposure at 5 mg/kg (lung burden of 5 mg) was without significant adverse effects. In another intratracheal instillation study, Henderson et al. (1995) exposed F344 rats to TiO_2 at up to 0.75 mg/ g of lung and characterized responses by BAL-fluid and histopathological analyses for 6 months following exposure. No significant effects were detected after TiO_2 exposure.

Reproductive and Developmental Effects. No data are available on the reproductive and developmental effects of acute TiO_2 exposure.

Repeated Exposures

Pulmonary Effects. Thomson et al. (1988) exposed rats to fine TiO_2 particles at 101.5 mg/m^3 for 4hr per day for 4 days and examined responses by histopathological and BAL-fluid analyses and pulmonary-function tests 24 hr and 14 days after exposure. The mass median aerodynamic (MMAD) of the particles ranged from 1.39 to 1.60 μm and the standard geometric deviation ranged from 2.06 to 2.10 μm. TiO_2 exposure resulted in increased BAL-fluid neutrophils and protein at 24 hr but not 14 days after exposure. No remarkable changes in pulmonary function were observed in this study, nor were any adverse histopathological changes seen in the lungs or other tissues.

Driscoll et al. (1991) exposed rats via inhalation to fine TiO_2 particles at 50 mg/m^3 for 6 hr per day for 5 days and examined responses by analyses of BAL-fluid, alveolar macrophage-derived cytokine production, and histopathological changes in the lungs for up to 63 days after exposure. The MMAD of the particles was 1.0 μm and the geometric standard deviation was 2.6 μm. Lung TiO_2 burdens of 1.8 mg were present at the end of the 5-day exposure. No changes in BAL-fluid indicators of injury and inflammation or in alveolar macrophage-derived cytokine production were observed. Histopathological examination revealed the presence of particle-containing macrophages but no adverse responses.

Henderson et al. (1995) exposed rats to fine TiO_2 particles via inhalation for 6 hr per day, 5 days per week for 4 weeks to 0.1 (MMAD was not determined), 1.0 (MMAD = 1.2 μm), or 10.0 (MMAD = 1.3 μm ± 0.01) mg/m^3. Responses were determined by BAL-fluid and histopathological analyses for up to 6 months following exposure. Lung TiO_2 burdens after the 4-week exposure were 4.4, 72, and 440 μg/g of lung. No adverse responses were observed in any TiO_2-exposed animals at any time point examined.

Warheit et al. (1996) exposed rats to fine TiO_2 particles at 5 and 250 mg/m^3 for 6 hr per day, 5 days per week for 4 weeks and examined inflammatory and proliferative responses in the lungs for up to 6 months following exposure. Marked and persistent lung inflammation and increased lung-cell proliferation were detected in rats exposed at 250 mg/m^3. Exposure at 5 mg/m^3 did not elicit any adverse effects.

Johnson and Wagner (1989) exposed rats to aerosols of TiO_2 at 10 mg/m^3 for 7 hr per day, 5 days per week for 12 weeks. Histopathological examination was performed on some animals at the end of the 12-week inhalation exposure and other animals were examined 12 months after exposure. The only remarkable responses detected were particle-containing alveolar macrophages and some pleural fibrosis.

TiO_2 particles are typically fine (i.e., diameter 0.1-2.5 μm); however, some specially manufactured TiO_2s are ultrafine (i.e., diameter <0.1 μm). Ferin et al. (1992), Oberdörster et al. (1994a,b) and Driscoll and Maurer (1991) examined the effects of ultrafine TiO_2 particles. Ferin et al. (1992) exposed rats to aerosols of ultrafine (diameter 0.02 μm) or fine (diameter 0.25 μm) TiO_2 at concentrations of 23.5 and 23.0 mg/m^3, respectively. Exposures were for 6 hr per day, 5 days per week for 12 weeks. The MMAD and the geometric standard deviation of the two aerosols were similar because when aerosols are generated, the materials form aggregates that behave as particles of similar size. At the end of exposure ultrafine- and fine-particle-exposed rats had similar lung burdens of TiO_2. BAL-fluid neutrophil numbers were increased to a significantly greater extent in rats exposed to the ultrafine particles compared with those exposed to the fine particles. Similar observations on ultrafine and fine TiO_2 particles were reported by Driscoll and Maurer (1991). Briefly, exposure of rats by intratracheal instillation of ultrafine TiO_2 particles at 10 mg/kg (lung burden of 1.8 mg) resulted in a marked and persistent inflammatory response in the lung. That response contrasted with the transient inflammation seen after exposure to fine TiO_2 particles. A greater increase in alveolar macrophage-derived inflammatory cytokines was produced by ultrafine TiO_2 particles than by fine particles, and a fibrotic response in the lung after ultrafine-particle exposure was not seen after fine-particle exposure. These studies indicate that ultrafine TiO_2 particles are more toxic in the lung than fine particles. A comparison of the effects of ultrafine versus fine TiO_2 particles on lung inflammation and lung-clearance retardation indicates that those responses correlate with the surface area of the TiO_2 particles and not the mass (Oberdörster et al. 1994a,b). Ultrafine particles have a markedly greater surface area than fine particles and thus, are more active.

Lee et al. (1985a,b) exposed rats to fine TiO_2 particles at concentrations of 10, 50, and 250 mg/m^3 for 6 hr per day, 5 days per week for 24 months. The MMAD of the particles was 1.5 to 1.7 μm. Lung burdens of TiO_2 after 24 months of exposure were 32, 130, and 545 mg per lung in the 10-, 50-, and 250-mg/m^3 exposure groups, respectively (Lee et al. 1986). Although TiO_2 was found in nonrespiratory tissues, including the liver and spleen, adverse responses were confined to the respiratory tract and associated lymphatic tissues (Lee et al. 1985a). Animals exposed at 10 mg/m^3 had increased numbers of particle-laden lung macrophages and minimal alveolar epithelial-cell hyperplasia. Exposure at 50 or 250 mg/m^3 resulted in a dose-related increase in alveolar epithelial hyperplasia, alveolar proteinosis, and pulmonary fibrosis.

Muhle et al. (1991) exposed rats to fine TiO_2 particles at a concentration of 5 mg/m^3 for 6 hr per day, 5 days per week for 24 months. The MMAD of the particles was 1.1 μm and the geometric standard deviation was 1.6 μm. Responses were characterized by BAL-fluid, histopathological and lung-clearance analyses. The lung burden of TiO_2 at the end of exposure was 2.7 mg per lung. Slight and variable increases in BAL-fluid neutrophils and lymphocytes were detected, and no changes in

BAL-fluid biochemical indicators of injury or edema were seen. Histopathological examination revealed an increase in particle-laden macrophages in the lungs; no lung fibrosis or lung tumors were observed.

Heinrich et al. (1995) exposed rats and mice to ultrafine TiO_2 at an average concentration of 10 mg/m^3 for 18 hr per day, 5 days per week. The MMAD of the particles was 0.8 μm and the geometric standard deviation was 1.8 μm. The rats were exposed for 24 months and the mice were exposed for 13.5 months followed by 9.5 months in clean air. After 24 months of exposure, the rats had an average of 39.3 mg of TiO_2 per lung and exhibited significantly increased lung weights, lower body weights, and life-time shortening compared with clean-air control animals. BAL-fluid markers of lung injury and inflammation were increased after TiO_2 exposure. Histopathological examination revealed particle-laden macrophages, bronchoalveolar hyperplasia, and slight fibrosis after 6 months of exposure. After 24 months, slight-to-moderate fibrosis and bronchoalveolar hyperplasia were observed. In this study, similar histopathological effects were reported for rats exposed at a similar concentration of ultrafine carbon-black particles and diesel soot. Mice exposed to ultrafine TiO_2 particles had average lung burdens of 5.2 mg per lung at the end of 12 months of exposure. Increased mortality and lung weights were observed in the exposed mice. Histopathological results were not reported.

Reproductive and Developmental Effects. No data are available on the reproductive and developmental effects of TiO_2 exposure.

Carcinogenic Effects. Two studies observed an increased incidence of lung tumors in rats exposed by inhalation to TiO_2. Lee et al. (1985a, b) exposed rats to fine TiO_2 particles at concentrations of 10, 50, and 250 mg/m^3 for 6 hr per day, 5 days per week for 24 months. At 250-mg/m^3, a significant increase in lung adenomas and carcinomas was observed (17% of TiO_2-exposed rats versus 1% of controls). The incidence of carcinomas was most prominent in females (13 of 74 females versus 1 of 77 males). An increase in lung tumors was not observed at 10 and 50 mg/m^3.

Heinrich et al. (1995) reported an increased incidence of lung tumors in female rats (32% in females versus 0.5% in controls) exposed to ultrfine TiO_2 particles at an average concentration of 10 mg/m^3 for 18 hr per day, 5 days per week for 24 months followed by 6 months in clean air.

In this study, no increase in lung tumors was detected in mice exposed at a similar concentration of ultrafine TiO_2 for 13.5 months and held for an additional 9.5 months.

In a study by Muhle et al. (1991), no increase in lung tumors was observed in rats exposed to fine TiO_2 particles at 5 mg/m^3 for 6 hr per day, 5 days per week for 24 months.

The above described tumorigenic effects of TiO_2 in rats are not unique to TiO_2 (Driscoll 1996). Several studies have shown that chronic inhalation of poorly soluble, low-toxicity particles can result in lung adenomas and carcinomas in rats. Other particulate materials shown to cause these tumorigenic effects include diesel soot, carbon black, and talc (Mauderly et al. 1987; NTP 1993; Heinrich et al. 1995). As with TiO_2, the tumorigenic response to those materials appears to be unique to the rat because hamsters and mice do not respond to similar lung burdens of those poorly soluble particles by developing lung cancer (NTP 1993; Heinrich et al. 1995). The mechanisms underlying the tumorigenic response of rats to poorly soluble particles and the relevance of that response to humans are not known. Recent analyses of the chronic inhalation data with poorly soluble particles indicate that the rat lung-tumor response correlates with the log particle surface area of the particles in the lung (Oberdörster and Yu 1990; Driscoll 1996). The correlation with surface area suggests why Heinrich et al. (1995) observed increases in rat lung tumors at an exposure concentration and lung mass burden of ultrafine TiO_2 particles that were below the concentration and mass burden for fine TiO_2 particles that produced no rat lung tumors in a study by Lee et al. (1985).

Mutagenic and Genotoxic Effects. Gallagher et al. (1994) examined lungs for the presence of DNA adducts in rats exposed chronically to ultrafine TiO_2 (MMAD = 0.82 μm). DNA was isolated from rats exposed at an average concentration of 10 mg/m^3 for 18 hr per day, 5 days per week for 24 months. No increase in the number or types of DNA adducts was observed in TiO_2-exposed rats compared with clean-air controls. In contrast, a type of DNA adduct not observed in rat lungs chronically exposed to TiO_2 or carbon black was detected in the lungs of rats exposed to diesel soot.

Driscoll et al. (1997) exposed rats via intratracheal instillation to a fine TiO_2 at 10 or 100 mg/kg of body weight and characterized mutation in the hgprt gene of alveolar epithelial cells isolated from the animals 15 months after exposure. A slight but significant increase in hgprt mutation frequency was observed in alveolar epithelial cells from rats exposed to TiO_2 at 100 mg/kg. The effect of TiO_2 was remarkably less than that in rats exposed to the same concentrations of quartz or carbon black. In vitro exposure of epithelial cells to TiO_2 did not result in increased hgprt mutant frequency.

Effects on Host Defense. Gilmour et al. (1989a,b) examined the effects of TiO_2 inhalation on clearance of and immune responses to *Pasteurella haemolytica*. Mice were exposed to an aerosol of fine TiO_2 particles at 20 mg/m^3 for 20 hr per day for periods of up to 7 weeks. After varying periods of exposure, mice were challenged with an aerosol of *P. haemolytica*, and the clearance of the bacteria was determined or the ability of the mice to mount an immune response was evaluated. Exposure to TiO_2 for 2 weeks resulted in impaired lung clearance of *P. haemolytica*. The impaired clearance of *P. haemolytica* was observed in mice exposed to TiO_2 immediately before or after bacterial challenge. TiO_2 exposure depressed the responses of cells isolated from the mediastinal lymph nodes to bacterial antigen challenge in vitro.

Creutzenberg et al. (1990) and Heinrich et al. (1995) reported on clearance of tracer particles from the lungs of rats exposed chronically to an aerosol of ultrafine TiO_2 particles. Rats were exposed at an average concentration of 10 mg/m^3 for 18hr per day, 5 day per week for 18 months with 6 months recovery in clean air. After 3, 12, and 18 months of exposure, animals were exposed to iron oxide (diameter 0.35 μm) or latex (diameter 3.5 μm) tracer particles, and lung retention of the tracer material was determined. Exposure to TiO_2 resulted in a significant increase in tracer-particle retention at all times examined. Average half-times of iron oxide clearance ranged from 61 to 93 days in clean-air control rats and from 208 to 368 days in TiO_2-exposed rats. Clearance of the latex tracer particles was more variable; longer half-times were observed after 3 and 6 months exposure to TiO_2, and shorter half-times were observed thereafter. The reason for the variability in the latex clearance was suggested to be due to the more-proximal deposition of the large latex particles (diameter 3.5 μm), a deposition that resulted from altered architecture of the rats' lungs after long-term TiO_2 exposure.

Dermal and Ocular Exposures

Lethality

The approximate lethal dose of dermally applied TiO_2 (percent lethality unspecified) is reported to be $>10,000$ mg/kg in rabbits (Trochimowicz and Reinhardt 1988).

Skin and Eye Irritation

TiO_2 is reported to be neither a dermal irritatant nor a dermal sensitizer and evoked only mild irritation in the eye (Trochimowicz and Reinhardt 1988).

In Vitro Studies

TiO_2 was nonmutagenic in the Ames assay using *Salmonella typhimurium* strains TA100, TA1535, TA97, and TA98 with or without metabolic activation (Zeiger et al. 1988). TiO_2 was tested and found to be negative in the Ames assay, the mouse lymphoma assay, and the *Drosophila* assay for induction of sex-linked recessive lethal mutations (NCI 1979). In Chinese hamster ovary cells, TiO_2 was negative for sister chromatid exchange; the results were inconclusive for chromosomal aberrations (NCI 1979). In vitro exposure of rat lung epithelial cells to TiO_2 did not increase the frequency of mutation in the hgprt locus (Driscoll et al. 1997).

SUMMARY OF TOXICITY DATA

Noncancer Effects

Available data on dermal and ocular exposure are sparse; however, these data and the physical and chemical properties of TiO_2 indicate a very low order of activity by those routes. After inhalation of very high concentrations of TiO_2, particles have been found in nonrespiratory tissues; however, adverse effects have been confined to the respiratory tract and lung-associated lymphatic tissues (Lee et al. 1995b).

A single brief inhalation exposure of rats to fine TiO_2 particles at 2260 mg/m^3 has not been associated with adverse effects. In a well-described study by Hilaski et al. (1992), a single 30- min exposure to TiO_2 at average concentrations of 1,240 to 830 mg/m^3 produced no adverse effects. In the course of the 30-min exposure, rats experienced a 5-min average TiO_2 concentration of 2,260 mg/m^3 and a 15-min average of 1,506 mg/m^3. In addition to this acute exposure study, the 4-hr lethal concentration for 50% of the test animals (LC_{50}) in rats is >6,820 mg/m^3 (DuPont 1979; Trochimowicz and Reinhardt, 1988).

Repeated inhalation of fine TiO_2 particles at >20 mg/m^3 is reported to produce inflammation, fibrosis, and epithelial hyperplasia in rat lung (Thomson et al. 1988; Ferin and Oberdörster 1992; Oberdörster et al. 1994a,b; Warheit et al. 1996). Additional effects observed after inhalation of TiO_2 at >20 mg/m^3 are impairment in the clearance of both viable and nonviable particles and a decreased immune response to inhaled bacteria. Overall, subchronic (Henderson et al. 1995) and chronic inhalation studies (Lee et al. 1995a,b) in rats do not demonstrate adverse effects at concentrations of 10 mg/m^3 for chronic inhalation of fine TiO_2 particles.

Studies using ultrafine TiO_2 particles, indicate that these particles would be expected to have an effect at a lower concentration than fine particles in the rat lung (Driscoll et al. 1991; Ferin et al. 1992; Oberdörster et al. 1994a,b; Heinrich et al. 1995). The difference in activity of fine and ultrafine TiO_2 particles correlates with the difference in surface area of these particles in the lung (Oberdörster et al. 1994a,b; Driscoll 1996).

Carcinogenic Effects

Data show an increase in lung-tumor incidence in rats exposed to TiO_2. A significant increase in the incidence of lung adenomas and carcinomas was observed in rats exposed by inhalation to fine TiO_2 particles for 6 hr per day, 5 days per week for 24 months at a concentration of 250 mg/m^3 (Lee et al. 1985a,b). No increase in the incidence of lung tumors was reported at exposure concentrations of 10 mg/m^3 and 50 mg/m^3. In a study by Heinrich et al. (1995), female rats exposed by inhalation to ultrafine TiO$_2$ particles for 6 hr per day, 5 days per week for 24 months at a concentration of 10 mg/m^3 had an increased incidence of lung tumors. In addition to TiO$_2$, other poorly soluble, low-toxicity particles have been shown to increase the incidence of lung tumors in rats. Hamsters and mice do not have an increased incidence of lung tumors even when lung burdens of poorly soluble, low-toxicity particles are similar to those of rats. The relevance of the data to humans is not known.

PREVIOUS RECOMMENDED EXPOSURE LIMITS

The Occupational Safety and Health Administration (OSHA) established a permissible exposure limit (PEL) of 15 mg/m^3 for total particulate TiO$_2$ and 5 mg/m^3 for respirable particulate TiO$_2$ (U.S. Department of Labor 1997). The National Institute for Occupational Safety and Health (NIOSH) has not established a recommended exposure limit and has recommended that the TiO$_2$ concentration be maintained as low as possible because of its recognition as a potential occupational carcinogen (NIOSH 1996). The American Conference of Governmental Industrial Hygienists (ACGIH) established a Threshold Limit Value–time-weighted average (TLV-TWA) of 10 mg/m^3 for TiO$_2$ (ACGIH 1996).

SUBCOMMITTEE EVALUATION AND RECOMMENDATIONS

MILITARY EXPOSURES

Emergency Exposure Guidance Levels (EEGLs)

No acute toxicity information is available on human exposure to TiO$_2$. On the basis of animal studies, the potential for lethality from a single exposure to TiO$_2$ is minimal. The 4-hr LC$_{50}$ for TiO$_2$ in the rat is >6,820 mg/m^3 (DuPont 1979). Hilaski et al. (1992) reported that a 30-min exposure of rats to a TWA concentration of up to 1,240 mg/m^3 generated from an XM82 grenade had no adverse effects characterized by histopathological examination, BAL-fluid analysis, and pulmonary-function tests. Additional information relevant to recommending acute exposure limits for TiO_2 comes from the extensive subchronic and chronic inhalation data base on TiO_2 and several other poorly soluble, low-toxicity particles (e.g., talc, graphite, and carbon black). When lung burdens of poorly soluble, low-toxicity particles at > 1-2 mg/g of lung tissue are obtained in the rat, considerable evidence shows that there is an overloading of alveolar-clearance mechanisms and a consequent slowing of particle clearance (Morrow et al. 1991; Oberdörster 1994). Overloading of the alveolar particle-clearance mechanisms has been observed in all experimental animals tested to date, making it reasonable to assume that the phenomenon occurs in humans (Oberdörster 1994). Associated with the alterations in particle clearance in the rat is the development of pulmonary inflammation, fibrosis, and epithelial hyperplasia; exposures resulting in lung burdens below 1-2 mg/g of lung appear to have no without adverse effects (Morrow et al. 1991; Oberdörster 1994). Overloading of the alveolar particle-clearance mechanisms has been shown to be more closely related to the volume than the mass of particles in the lung (Morrow 1988; Morrow et al. 1991). In this respect, for TiO_2, the lung mass burden expected to cause particle-clearance changes and adverse lung effects would need to be increased to account for the high density of TiO_2 (Morrow et al. 1991).

The approach the subcommittee used to recommend EEGLs was to estimate the maximal acute inhalation exposure concentration that would result in a lung particle dose below that associated with impairment of particle clearance and adverse lung effects. On the basis of long-term inhalation studies in rodents, a lung dose of TiO_2 at <4 mg/g of lung would not be expected to alter alveolar particle clearance. The 4 mg/g lung value incorporates a correction for the average density (4 g/cm^3) of the two forms of TiO_2. Because the 4-mg/g lung dose is based on results of long-term inhalation studies, the subcommittee considered the possibility that the response could be greater to a dose of 4 mg/g of lung delivered by acute exposure (high-dose rate) than by chronic exposure (low-dose rate). In considering the possibility for dose-rate effects, the subcommittee noted that in intratracheal instillation studies, bolus doses of approximately 0.7 to 2 mg of TiO_2 have been delivered to the lungs of rats and resulted in either no effects or transient pulmonary inflammation (Driscoll et al. 1990a,b; Henderson et al. 1995). These studies indicate that even at the high-dose rates associated with intratracheal instillation exposure, high lung burdens of TiO_2 have minimal effects on the lung. Exposure to TiO_2 delivered by inhalation would not be expected to lead to a greater response compared with exposure delivered by instillation (Driscoll et al. 1991). Still, to address the potential for greater response to high-dose rate in acute inhalation exposures, the subcommittee reduced the 4-mg/g lung dose by a factor of 10 and estimated the maximal 15-min, 1-hr, and 6-hr exposure concentrations that would result in a lung dose of <0.4 mg/g of lung. Making reasonable assumptions for minute volume (43 liters/min, heavy work activity; Diem and Lentner 1970), deep-lung particle deposition (35% of respirable particles; Stahlhofen et al. 1989), and lung weight (1,000 g; Diem and Lentner 1970), a 15-min exposure to approximately 1,770 mg/m^3 was calculated to result in a deep-lung dose of <0.4 mg/g of lung[2] and have no toxic effects. Studies that support that calculation are the 4-hr inhalation study in rats demonstrating no mortality after exposure to 6,820 mg/m^3 (DuPont 1979), the intratracheal instillation studies demonstrating that bolus administration of 0.7 to 2 mg of TiO_2 into rat lungs produced either no adverse effects or transient pulmonary inflammation (Driscoll et al. 1990b; Henderson et al. 1995), an acute inhalation study demonstrating no adverse effects in rats after a 30-min exposure at up to 1,240 mg/m^3 (Hilaski et al. 1992), and a chronic inhalation study demonstrating minimal adverse lung effects in rats after exposure to a concentration of 250 mg/m^3, which resulted in a lung dose of >10 mg/g of lung (Lee et al. 1986). Based on the above considerations, the subcommittee recommends 1,800 mg/m^3 (rounded from 1,770 mg/m^3) as the 15-min EEGL for TiO_2. Extrapolating from the 15-min EEGL, a 1-hr EEGL of 450 mg/m^3 for respirable dust and a 6-hr EEGL of 75 mg/m^3 for respirable dust are recommended.

---

[2]Calculation of 15-min exposure concentration in humans resulting in a lung TiO_2 dose of 0.4 mg/g of lung.

Exposure concentration = [(0.4 mg/g of lung)(1,000 g/lung)]/
 [(43 L/min)(15 min)(0.35)(m^3/1,000)].

The above recommended exposure limits are for fine-sized TiO_2 particles, which are defined as particles with a diameter of 0.1-2.5 μm. Although it is expected that exposures to TiO_2 from its use in military smokes and obscurants will be to fine particles, the subcommittee recognizes that there are specialized forms of TiO_2 manufactured as ultrafine materials. Studies in animals indicate that ultrafine TiO_2 particles (diameter <0.1 μm) have a greater toxicity in the rat lung than do fine TiO_2 particles (Driscoll et al. 1991; Ferin et al. 1992; Oberdörster et al.1994a,b; Heinrich et al. 1995). A comparison of the lung response to fine and ultrafine TiO_2 particles indicate that lung inflammation and tumorigenic responses to these materials correlate with the surface area of the particles (Oberdörster and Yu 1990; Driscoll 1996). The surface area for ultrafine TiO_2 particles is reported to be approximately 50 m^2/g and that of fine TiO_2 particles is typically 6-8 m^2/g (Ferin et al. 1992; Oberdörster et al. 1994a,b; Driscoll 1996), an approximate 8-fold difference in surface area. If exposures to ultrafine TiO_2 particles are of concern, an adjustment in the EEGLs based on particle surface-area differences should be considered and the EEGLs reduced by an 8-fold factor.

Repeated Exposure Guidance Level (REGL)

In considering an appropriate exposure limit for repeated exposure to TiO_2, the subcommittee noted that chronic inhalation to high concentrations of fine or ultrafine TiO_2 particles is shown to result in pulmonary inflammation, fibrosis, and lung tumors in rats (Lee et al. 1985a,b; Heinrich et al. 1995). However, an epidemiological study did not find a increased cancer risk in TiO_2-exposed humans (Chen and Fayerweather 1988). As discussed above, TiO_2 is not unique among particles in producing inflammatory, fibrotic, and neoplastic responses in rats but is one of a number of low-toxicity, poor-solubility particles shown to produce these adverse effects at high concentrations (Oberdörster 1994). In recommending an appropriate exposure guideline for repeated inhalation exposure to such particles, several factors that are important include the likely mechanism for the lung-tumor response and the dosimetry of the material.

Increasing evidence indicates that persistent inflammation and increased cell proliferation play a key role in the development of lung tumors in rats after chronic inhalation at high concentrations of poorly soluble, nongenotoxic particles, such as TiO_2 (Oberdörster 1994; Driscoll 1996). The release of oxidants by inflammatory cells provides a mechanism by which damage to DNA can occur after exposure to nongenotoxic particles (Ames 1983; Weitzman et al. 1985). Enhanced cell proliferation can increase the probability that a spontaneous genetic error will occur during normal cell division, resulting in mutation (Cohen and Ellwein 1991). In the presence of added oxidative stress, cell proliferation can be expected to increase the likelihood that spontaneously occurring and oxidant-induced mutations will be fixed in a proliferating cell and clonally expanded. The result is an increase in the cell population at risk for genetic changes relevant to tumorigenesis. From the perspective of recommending exposure guidelines, an important factor implicit in that mechanism is that a threshold exists, and exposure concentrations that do not produce persistent inflammation and increased cell proliferation will not carry an increased risk of lung tumors.

The relation between exposure concentration and lung dose of poorly soluble particles is a key factor in recommending acceptable exposure guidelines. Several studies in rats indicate that when lung burdens of poorly soluble particles are >1-2 mg/g of lung, alveolar particle-clearance mechanisms are overloaded and particle retention is increased (Morrow et al. 1991; Oberdörster 1994). Consequently, after repeated exposure to high concentrations, accumulation of particles in the lung is disproportionate to that occurring at lower concentrations. Associated with the accumulation of lung particle burdens of >1-2 mg/g of lung in the rat is the development of adverse lung effects, including inflammation, fibrosis, and epithelial hyperplasia. Thus, the data base on chronic inhalation of low-toxicity, poorly soluble particles by rats indicates that lung burdens of >1-2 mg/g of lung carry an increased risk of adverse lung effects that might contribute to the development of lung cancer in the rat (Morrow 1992). It is further noted that the overloading of lung particle-clearance mechanisms in the rat is more closely related to the volume than the mass of particles in the lung (Morrow 1992).

Because the effects of high lung doses on particle clearance have been observed in all experimental animals tested to date, it reasonable to assume that the phenomenon occurs in humans (Oberdörster 1994).

Thus, the subcommittee considered that a reasonable approach to recommending a REGL for TiO_2 was to estimate the exposure concentration for humans that would not overload alveolar clearance mechanisms and result in the adverse effects seen in the rat studies. That approach for developing occupational exposure limits has been described by Morrow et al. (1991) and Oberdörster (1994). Making reasonable assumptions for minute ventilation (22 liters/min, moderate work activity; Raabe 1979), the proportion of respirable particles depositing in the deep lung (35%; Stahlhofen et al. 1989), lung weight (1,000 g; Diem and Lentner 1970), and particle-clearance half-time (500 days; Bohning et al. 1982) for humans and accounting for the density of TiO_2, the subcommittee estimated that chronic inhalation of poorly soluble particles at 2 mg/m³ for 8 hr per day, 5 days per week is the maximal exposure concentration that would result in a lung burden of <4 mg/g of lung.[3] The calculation takes into account sources of uncertainty in lung dosimetry between rodents and humans and therefore, an additional uncertainty factor extrapolating from animals to humans was not applied. Additionally, the uncertainty factor was reduced to 1 because the rat is a more sensitive species than the human to the effects of poorly soluble particles in the lung. A REGL of 2 mg/m³ for fine TiO_2 particles is supported by the epidemiological study of Chen and Fayerweather (1988) who found no increase in malignant and nonmalignant respiratory disease in workers with TWA exposures to TiO_2 estimated at up to 20 mg/m³.

The REGL recommended by the subcommittee is less than the 5-mg/m³ OSHA PEL for respirable TiO_2. However, it should be noted that the approach used by the subcommittee (i.e., estimating the concentration that would not overload alveolar particle-clearance mechanisms) to recommend the REGL is the same as that used by the ACGIH TLV committee in developing the recently updated TLV for particles not otherwise classified (PNOC) (ACGIH 1996).

---

[3]Calculation of exposure concentration (8 hr per day × 5 days per week) resulting in a lung TiO_2 dose of 4 mg/g of lung.

Exposure Concentration = [(4 mg/g of lung)] ÷ {[(22 L/min)(60 min) (8 h/d)(5 d/7 d)(m³/1,000)(0.35) (lung/1,000 g)] ÷ [ln 2/500 d]}.

As discussed for EEGLs, ultrafine TiO_2 particles appear to be more toxic than fine TiO_2 particles in the rat lung. Because of the positive correlation between surface-area dose and biological activity of fine and ultrafine TiO_2 particles, an adjustment in the REGL based on particle surface-area differences should be considered if exposures to ultrafine TiO_2 particles are of concern (Oberdörster et al. 1992; Driscoll 1996).

PUBLIC EXPOSURES

Short-Term Public Emergency Guidance Levels (SPEGLs)

To recommend a SPEGL for TiO_2, the subcommittee divided the EEGLs by an uncertainty factor of 10 to account for the potentially greater range in susceptibilities of the general public compared with healthy military personnel. Thus, the recommended 15-min SPEGL is 180 mg/m^3, the 1-hr SPEGL is 45 mg/m^3, and the 6-hr SPEGL is 7.5 mg/m^3 for respirable TiO_2 dust. As with the EEGLs and the REGL, in the event that exposures are to ultrafine TiO_2 particles, an adjustment in the SPEGLs based on particle surface-area differences can be considered.

Repeated Public Exposure Guidance Level (RPEGL)

The subcommittee recommends that the RPEGL be derived by incorporating an uncertainty factor of 10 into the REGL to account for the potentially greater range in susceptibilities of the general public compared with healthy military personnel. Therefore, the recommended RPEGL for TiO_2 is 0.2 mg/m^3 for respirable TiO_2 dust.

SUMMARY OF SUBCOMMITTEE RECOMMENDATIONS

Table 4-2 summarizes the subcommittee's recommendations for exposure limits for military personnel exposed to TiO_2. The subcommittee's recommendations for exposure limits for the general public are listed in Table 4-3.

TABLE 4-2 EEGLs and REGL for Fine TiO_2-Particle Smoke for Military Personnel

Exposure Guideline	Exposure Duration	Exposure Guidance Level (mg/m³ respirable dust)
EEGL	15 min	1800
	1 hr	450
	6 hr	75
REGL	8 hr/d, 5 d/wk	2

Abbreviations: EEGL, emergency exposure guidance level; REGL, repeated exposure guidance level.

TABLE 4-3 SPEGL and RPEGL for Fine TiO_2-Particle Smoke at the Boundaries Military-Training Facilities

Exposure Guideline	Exposure Duration	Exposure Guidance Level (mg/m³ respirable dust)
SPEGL	15 min	180
	1 hr	45
	6 hr	7.5
RPEGL	8 hr/d, 5 d/wk	0.2

Abbreviations: SPEGL, short-term public emergency guidance levels; RPEGL, repeated public exposure guidance level.

RESEARCH NEEDS

In the event that the military considers the use of ultrafine TiO_2 particles in smokes and obscurants, research to clearly define the differences in the toxic potential of the ultrafine particles and fine particles should be considered.

REFERENCES

ACGIH (American Conference of Governmental Industrial Hygienists). 1996. Guide to Occupational Exposure Values—1996. American Conference of Governmental Industrial Hygienists, Cincinnati, Ohio.

Ames, B.N. 1983. Dietary carcinogens and anticarcinogens. Oxygen radicals and degenerative diseases. Science 221:1256-1264.

Bellmann, B., H. Muhle, O. Creutzenberg, C. Dasenbrock, R. Kilpper, J.C. MacKenzie, P. Morrow, and R. Mermelstein. 1991. Lung clearance and retention of toner, utilizing a tracer technique, during chronic inhalation exposure in rats. Fundam. Appl. Toxicol. 17:300-313.

Bohning, D.E., H.L. Atkins, and S.H. Cohn. 1982. Long-term particle clearance in man: normal and impaired. Ann. Occup. Hyg. 26:259-271.

Chen, J.L., and W.E. Fayerweather. 1988. Epidemiologic study of workers exposed to titanium dioxide. J. Occup. Med. 30:937-942.

Cohen, S.M., and L.B. Ellwein. 1991. Genetic errors, cell proliferation, and carcinogenesis. Cancer Res. 51:6493-64505.

Creutzenberg, O., B. Bellmann, U. Heinrich, R. Fuhst, W. Kolch, and H. Muhle. 1990. Clearance and retention of inhaled diesel exhaust particles, carbon black, and titanium dioxide in rats at lung overload conditions. J. Aerosol Sci. 21(Suppl. 1):S455-S458.

Diem, K., and C. Lentner, eds. 1970. Documenta Geigy: Scientific Tables, 7th Ed. Basle, Switzerland: Ciba-Geigy.

Driscoll, K.E. 1996. Role of inflammation in the development of rat lung tumors in response to chronic particle exposure. Inhalation Toxicol. 8:139-153.

Driscoll, K.E., and J.K. Maurer. 1991. Cytokine and growth factor release by alveolar macrophages: Potential biomarkers of pulmonary toxicity. Toxicol. Pathol. 19(4 Pt. 1):398-405.

Driscoll, K.E., R.C. Lindenschmidt, J.K. Maurer, J.M. Higgins, and G. Ridder. 1990a. Pulmonary response to silica or titanium dioxide: Inflammatory cells, alveolar macrophage-derived cytokines, and histopathology. Am. J. Respir. Cell Mol. Biol. 2:381-390.

Driscoll, K.E., J.K. Maurer, R.C. Lindenschmidt, D. Romberger, S.I. Rennard, and L. Crosby. 1990b. Respiratory tract responses to dust: Relationships between dust burden, lung injury, alveolar macrophage fibronectin release, and the development of pulmonary fibrosis. Toxicol. Appl. Pharmacol. 106:88-101.

Driscoll, K.E., R.C. Lindenschmidt, J.K. Maurer, L. Perkins, M. Perkins, and J. Higgins. 1991. Pulmonary response to inhaled silica or titanium dioxide. Toxicol. Appl. Pharmacol. 111:201-210.

Driscoll, K.E., L.C. Deyo, J.M. Carter, B.W. Howard, D.G. Hassenbein, and T.A. Bertram. 1997. Effects of particle exposure and particle-elicited inflammatory cells on mutation in rat alveolar epithelial cells. Carcinogenesis 18:423-430.

DuPont. 1979. Inhalation, Approximate Lethal Concentration—Titanium Diox-

ide. Rep. No. 77-79. E.I. du Pont de Nemours & Company, Wilmington, Del.
Ferin, J., and G. Oberdörster. 1985. Biological effects and toxicity assessment of titanium dioxides: Anatase and rutile. Am. Ind. Hyg. Assoc. J. 46:69-72.
Ferin, J., G. Oberdörster, and D.P. Penney. 1992. Pulmonary retention of ultrafine and fine particles in rats. Am. J. Respir. Cell Mol. Biol. 6:535-542.
Gallagher, J., U. Heinrich, M. George, L. Hendee, D.H. Phillips, and J. Lewtas. 1994. Formation of DNA adducts in rat lung following chronic inhalation of diesel emissions, carbon black and titanium dioxide particles. Carcinogenesis 15:1291-1299.
Gilmour, M.I., F.G.R. Taylor, and C.M. Wathes. 1989a. Pulmonary clearance of *Pasteurella haemolytica* and immune responses in mice following exposure to titanium dioxide. Environ. Res. 50:184-194.
Gilmour, M.I., F.G.R. Taylor, A Baskerville, and C.M. Wathes. 1989b. The effect of titanium dioxide inhalation on the pulmonary clearance of *Pasteurella haemolytica* in the mouse. Environ. Res. 50:157-172.
Heinrich, U., R. Fuhst, S. Rittinghausen, O. Creutzenberg, B. Bellmann, W. Koch, and K. Levsen. 1995. Chronic inhalation exposure of Wistar rats and two different strains of mice to diesel engine exhaust, carbon black, and titanium dioxide. Inhalation Toxicol. 7:533-556.
Henderson, R.F., K.E. Driscoll, J.R. Harkema, R.C. Lindenschmidt, I.-Y. Chang, K.R. Maples, and E.B. Barr. 1995. A comparison of the inflammatory response of the lung to inhaled versus instilled particles in F344 rats. Fundam. Appl. Toxicol. 24:183-197.
Hilaski, R.J., J.D. Bergmann, J.C. Carpin, W.T. Muse, Jr., and S.A. Thomson. 1992. Acute Inhalation Toxicity Effects of Explosively Disseminated—XM82 Grenade—Titanium Dioxide. CRDEC-TR-363. Chemical Research, Development and Engineering Center, U.S. Army Armament, Munitions and Chemical Command, Aberdeen Proving Ground, Edgewood, Md.
Johnson, N.F., and J.C. Wagner. 1989. Pleural and Parenchymal Responses in Rats to Short-Term Inhalation Exposures to Erionite, Crocidolite, Chrysotile, Silica and Titanium Dioxide. Pp. 157-164 in Effects of Mineral Dusts on Cells. NATO ASI Series, Vol. H30. B.T. Mossman and R.O. Bégin, eds. Berlin: Springler-Verlag.
Karlsson, N., G. Cassel, I. Fängmark, and F. Bergman. 1985. The inhalation toxicity of screening smokes [abstract]. J. Toxicol. Clin. Toxicol. 23(4-6):463.
Lee, K.P., H.J. Trochimowicz, and C.F. Reinhardt. 1985a. Transmigration of titanium dioxide (TiO_2) particles in rats after inhalation exposure. Exp. Mol. Pathol. 42:331-343.
Lee, K.P., H.J. Trochimowicz, and C.F. Reinhardt. 1985b. Pulmonary response

of rats exposed to Titanium Dioxide (TiO$_2$) by inhalation for two years. Toxicol. Appl. Pharmacol. 79:179-192.

Lee, K.P., N.W. Henry III, H.J. Trochimowicz, and C.F. Reinhardt. 1986. Pulmonary response to impaired lung clearance in rats following excessive titanium dioxide dust deposition. Environ. Res. 41:144-167.

Li, K., B. Keeling, and A. Churg. 1996. Mineral dusts cause elastin and collagen breakdown in the rat lung: A potential mechanism of dust-induced emphysema. Am. J. Respir. Crit. Care Med. 153:644-649.

Mauderly, J.L., R.K. Jones, W.C. Griffith, R.F. Henderson, and R.O. McClellan. 1987. Diesel exhaust is a pulmonary carcinogen in rats exposed chronically by inhalation. Fundam. Appl. Toxicol. 9:208-221.

Morrow, P.E. 1988. Possible mechanisms to explain dust overloading of the lungs. Fundam. Appl. Toxicol. 10:369-384.

Morrow, P.E., H. Muhle, and R. Mermelstein. 1991. Chronic inhalation study findings as a basis for proposing a new occupational dust exposure limit. J. Am. Col. Toxicol. 10:279-289.

Muhle, H., B. Bellmann, O. Creutzenberg, C. Dasenbrock, H. Ernst, R. Kilpper, J.C. MacKenzie, P. Morrow, U. Mohr, S. Takenaka, and R. Mermelstein. 1991. Pulmonary response to toner upon chronic inhalation exposure in rats. Fundam. Appl. Toxicol. 17:280-299.

NCI (National Cancer Institute). 1979. Bioassay of Titanium Dioxide for Possible Carcinogenicity. NCI Carcinog. Tech. Rep. Ser. Vol. 97. National Cancer Institute, Bethesda, Md.

NIOSH (National Institute for Occupational Safety and Health). 1996. Titanium dioxide. In Documentation for Immediately Dangerous to Life or Health. Online. Entry last updated Aug. 16, 1996. Available: http://www.cdc.gov/niosh/idlh/13463677.html. U.S. Department of Health and Human Services, National Institute for Occupational Safety and Health, Division of Standards Development and Technology Transfer, Cincinnati, Ohio.

NTP (National Toxicology Program). 1993. TR-421 Toxicology and Carcinogenesis Studies of Talc (CAS No. 14807-96-6) (Non-Asbestiform) in F344/N Rats and B6C3F$_1$ Mice (Inhalation Studies). TR-421. National Institute of Environmental Health Sciences, National Toxicology Program, Research Triangle Park, N.C. Available from NTIS, Springfield, Va., Doc. No. PB94-215985.

Oberdörster, G. 1994. Extrapolation of results from animal inhalation studies with particles to humans. Pp. 57-73 in Toxic and Carcinogenic Effects of Solid Particles in the Respiratory Tract, D.L. Dungworth, J.L. Mauderly, and G. Oberdörster, eds. Washington, D.C.: ILSI Press.

Oberdörster, G., and C.P. Yu. 1990. The carcinogenic potential of inhaled

diesel exhaust: A particle effect? J. Aerosol Sci. 21(Suppl. 1):397-401.
Oberdörster, G., J. Ferin, R. Gelein, S.C. Soderholm, and J. Finkelstein. 1992. Role of the alveolar macrophage in lung injury: Studies with ultrafine particles. Environ. Health Perspect. 97:193-199.
Oberdörster, G., J. Ferin, and B.E. Lehnert. 1994a. Correlation between particle size, in vivo particle persistence, and lung injury. Environ. Health Perspect. 102(Suppl. 5):173-179.
Oberdörster, G., J. Ferin, S. Soderholm, R. Gelein, C. Cox, R. Baggs, and P.E. Morrow. 1994b. Increased pulmonary toxicity of inhaled ultrafine particles: Due to lung overload alone? Pp. 295-302 in Proceedings of Seventh International Symposium on Inhaled Particles, J. Dogston and R.I. McCallum, eds. Oxford, U.K.: Pergamon.
Raabe, O.G. 1979. Deposition and Clearance of Inhaled Aerosols. UCD-472-503. Laboratory for Energy-Related Health Research, University of California at Davis for the U.S. Department of Energy, Washington, D.C.
Stahlhofen, W., G. Rudolf, and A.C. James. 1989. Intercomparison of experimental regional aerosol deposition data. J. Aerosol Med. 2(3):285-308.
Thomson, S.A., J.D. Bergmann, D.C. Burnett, J.C. Carpin, C.L. Crouse, R.J. Hilaski, B. Infiesto, and E. Lawrence-Beckett. 1988. Comparative Inhalation Screen of Titanium Dioxide and Graphite Dusts. CRDEC-TR-88161. Chemical Research, Development and Engineering Center, U.S. Army Armament, Munitions and Chemical Command, Aberdeen Proving Ground, Edgewood, Md.
Thomson, S.A., D.C. Burnett, J.C. Carpin, J.D. Bergmann, and R.J. Hilaaki. 1990. Comparative Inhalation Hazards of Titanium Dioxide, Synthetic and Natural Graphite. Proceedings of the Seventh International Pneumoconioses Conference. Part 2. U.S. Department of Health and Human Services, Centers for Disease Control and Prevention, Atlanta, Ga.
Trochimowicz, H.J., K.P. Lee, and C.F. Reinhardt. 1988. Chronic inhalation exposure of rats to titanium dioxide dust. J. Appl. Toxicol. 8:383-385.
U.S. Department of Labor. 1997. Occupational Safety and Health Standards. Air Contaminants. Code of Federal Regulations, Title 29, Part 1910, Section 1910.1000. Washington, D.C.: U.S. Government Printing Office.
Warheit, D.B., J.F. Hansen, I.S. Yuen, S.I. Snajdr, and M.A. Hartsky. 1996. Prolonged cellular proliferation and pulmonary inflammation are produced in rats inhaling high concentrations of low-toxicity insoluble particulates: Comparisons with cytotoxic dusts. Inhalation Toxicol. 8(Suppl.):155-167.
Yeh, H.C., M.B. Snipes, A.F. Eidson, and C.H. Hobbs. 1990. Comparative evaluation of nose-only versus whole-body inhalation exposures for rats—Aerosol characteristics and lung deposition. Inhalation Toxicol. 2:205-221.

Zeiger, E., B. Anderson, S. Haworth, T. Lawlor, and K. Mortelmans. 1988. *Salmonella* mutagenicity tests: IV. Results from the testing of 300 chemicals. Environ. Mol. Mutagen. 11(Suppl.12):1-158.

Zeng, L., Z. Zheng, and S. Zhang. 1989. Pathogenic effects of titanium dioxide dust on the lung of dogs—A histopathological and ultrastructural study [in Chinese]. J. West China Univ. Med. Sci. 20:88-91.

5

Graphite Smoke

BACKGROUND INFORMATION

MILITARY APPLICATIONS

THE MILITARY uses graphite flakes to block electromagnetic waves that the enemy might detect and use to target troops in the field. Graphite flakes provide the military with protection beyond the visual and infrared spectrums (Driver 1993).

The military releases the graphite flakes into the environment from ground-based systems that mechanically disperse bulk powders into the atmosphere (Lundy and Eaton 1994). The powder is used directly or compressed into small pellets to improve handling and delivery to the air ejector of smoke generators. Because the aerodynamic sizes of air-dispersed flakes are small (<20 micrometers (μm)), there is little near-source surface deposition of particles. Near-source deposition can be substantial if the air ejector is oriented at the ground or nearly parallel to the ground. The long-range downwind patterns resulting from dispersion and deposition depend on local meteorological conditions.

PHYSICAL AND CHEMICAL PROPERTIES

Graphite is a soft, crystalline form of carbon that occurs naturally and is also synthesized for various uses (ACGIH 1991). Natural graphite usually is associated with such impurities as quartz, mica, iron oxide, and granite (Hanoa 1983; Taylor 1985). Synthetic graphite is produced by heating a mixture of petroleum coke or coal, a binder (usually coal tar pitch), a petroleum-based oil to facilitate extrusion, and in some cases, anthracite coal (Watson et al. 1959; Zahorski 1960; Zahorski et al. 1975; Taylor 1985; Petsonk et al. 1988). The characteristics of the synthetic graphite depend on many factors, including the quality of the coke and binder, the degree of orientation of the particles during extrusion, and the temperature and time of processing (Taylor 1985).

Only synthetic graphite is used as an obscurant by the U.S. military. When it was classified, it was called EA 5768. The military generates the graphite smokes from one of two synthetic graphite powders, Micro-260 (manufactured by Asbury Graphite Mills, Inc.) and KS-2 (manufactured by Dixon Ticonderoga Company). Those graphites are composed of platelets or flakes of various sizes. In a recent study of the particle size of graphite from an XM56 generator, the mass median diameters for Types 1 and 2 reprocessed infrared material were 3-5 μm and 47-106 μm, respectively (Guelta et al. 1993). The chemical composition of the bulk powders is predominantly carbon, with trace impurities totaling <1% of the total weight. The trace impurities include small quantities of silica, aluminum, iron, calcium, titanium, and magnesium (Lundy and Eaton 1994).

TOXICOKINETICS

No data on the toxicokinetics of graphite were found. It can be assumed, as discussed by Driver et al. (1993), that inhaled particles, deposited in the deep lung, can be removed through phagocytosis by macrophages or by direct movement of the particles into the blood or the lymphatic tissues.

TOXICITY SUMMARY

The term "graphite" is used throughout this chapter to refer to graphite in the form of platelets, flakes, or dusts. The different forms do not appear to be associated with different toxicities. A review of the potential environmental and health effects of graphite flakes is available (Driver et al. 1993).

EFFECTS IN HUMANS

Graphite pneumoconiosis is a well-recognized pulmonary lesion that is found in workers involved in the mining and processing of graphite. Lungs diseased by graphite inhalation manifest characteristic pathological features. In general, they are diffusely discolored, ranging in appearance from gray to black, and are described as resembling "sponges that have been soaked in ink" (Pendergrass et al. 1968). Microscopically, the lungs show a granulomatous reaction with areas of interstitial fibrosis, perifocal emphysema, necrosis, and severe vascular sclerosis (Jaffé 1951; Gaensler et al. 1966). Occasionally, widespread tissue necrosis results in the formation of large cavities that become filled by an oily black liquid (Dunner and Bagnall 1946). Nonasbestos bodies with a black graphite core, superficially resembling the characteristic iron-containing bodies seen after asbestos exposure, are dispersed in the lungs and, occasionally, are found in the sputum (Mazzucchelli et al. 1996). Hanoa (1983) analyzed over 600 cases of graphite pneumoconiosis published in the literature. The author concluded that graphite pneumoconiosis is probably a mixed-dust type of lung reaction in most cases. Although analytically pure graphite cannot be excluded as causing chronic lung lesions, most investigators (Hanoa 1983; Zahorski et al. 1975) believe that it is the silica contamination that contributes to the development of fibrosis. That view is also shared by earlier investigators (Gloyne et al. 1949; Harding and Oliver 1949), although some cases have been described in which pneumoconiosis developed following exposure to synthetic graphite without any substantial silica content (Lister and Wimborne 1972). Two surveys were conducted in 1972 and 1989 in the graphite industry in Sri Lanka (Ranasinha and Uragoda 1972; Uragoda 1989). In 1972, 22.7% of the examined workers had chest x-ray signs suggestive of pneumoconiosis; in 1989, only 3.4% of the workers had such signs. However, the surveys were conducted in two different mines, and the conclusion that improved control measures were solely responsible for the decrease in disease incidence is tentative.

EFFECTS IN ANIMALS

Inhalation Exposures

Acute inhalation of graphite elicited only a transient inflammatory reaction in the respiratory tract of experimental animals. Anderson et al. (1989) exposed male Fischer 344 (F344) rats for a single 4-hr period to synthetic graphite (Asbury Micro-266) at 1 to 500 milligrams per cubic meter (mg/m^3). An increase in polymorphonuclear leukocytes in the bronchoalveolar lavage (BAL) fluid was observed 24 hr after exposure only in the animals exposed to a graphite concentration of 500 mg/m^3. After 7 days, the signs of inflammation had disappeared. Concentrations of 100 mg/m^3 or less failed to elicit any inflammatory response. Alveolar macrophages were activated only for 1 day following exposure and only at 500 mg/m^3 (Anderson et al. 1989).

The acute inhalation toxicity of graphite type 6353 and Microfyne graphite was determined in groups of five male and five female rats exposed for 1 hr at an airborne concentration of 23.7 milligrams per liter (mg/L) or 23,700 mg/m^3 (particle diameter <10 μm). The animals were observed for 2 weeks. Detailed histological or biochemical evaluations were not conducted; no gross signs of toxicity were seen (Manthei et al. 1980).

Repeated exposures of experimental animals to graphite appear to produce minimal effects. Thomson et al. (1988) exposed male F344 rats in whole-body chambers to synthetic (Asbury Micro-260, >1% silica) or natural graphite (Asbury Micro-650, 1.86% silica) at 100 mg/m^3 for 4 days, 4 hr per day. Mass median aerodynamic diameters (MMAD) of the particles ranged from 2.2 to 2.4 μm (geometric standard deviation 2.53 to 2.60 μm). Pulmonary function tests conducted 24 hr after exposure revealed a transient decrease in pulmonary resistance in rats exposed to synthetic graphite, although that observation was not considered biologically significant. Analysis of BAL fluid from the lungs of exposed animals 24 hr after exposure showed increased protein content (not statistically significant) and increased activity of alkaline phosphatase and β-glucuronidase for both graphite materials and increased lactate dehydrogenase activity for natural graphite. After 2 weeks, all biochemical changes were reversed. Both graphite materials also produced an increased influx of polymorphonuclear leukocytes into BAL fluid (up to 145% of control values); that change was also reversed after 2 weeks. Pigment in alveolar macrophages and free pigment in the alveoli were found 24 hr after exposure; after 2 weeks, no free pigment was seen. Only minuscule foci of alveolar type II cell hyperplasia were found in the peripheral lung tissue of a few animals. The authors concluded that inhalation of both materials produced minimal adverse effects.

Longer exposure durations were studied in another experiment

(Thomson et al. 1987). Male and female F344 rats were exposed to Micro-260. Chamber concentrations were 1, 10, and 100 mg/m^3, and exposures lasted for 2 hr per day, 5 days per week, for 2 weeks. MMADs followed by geometric standard deviations of the particles were 3.40 μm (1.90 μm), 2.90 μm (2.19 μm) and 1.78 μm (2.88 μm), respectively. Three days after exposure, increases in BAL-fluid protein and enzyme activity (lactate dehydrogenase, alkaline phosphatase, and β-glucuronidase) were observed only in the highest-exposure group and were higher than those observed after a single exposure. All changes disappeared 3 months after exposure. Similarly, increased cell content was found in the BAL fluid, most pronounced and significant in the highest-exposure group, and changes diminished with time. There was no evidence that graphite could increase the pulmonary content of hydroxyproline, a biochemical indicator of fibrosis. Hyperplasia of type II alveolar cells, labeled adenomatous hyperplasia, and histiocytosis were seen only in the highest-exposure group. Graphite material persisted in alveolar macrophages for 3 months. Overall, changes induced by the highest graphite concentration (100 mg/m^3) were considered mild and slowly reversible.

An extensive inhalation study with pure graphite or mixtures of graphite and fog oil was carried out in rats (Aranyi et al. 1992). All exposures were administered for 4 hr per day, 4 days per week. In a 4-week study, the rats were exposed to either graphite alone at 200 mg/m^3, fog oil alone at 500 or 1,000 mg/m^3, or a mixture of fog oil and graphite. Particle sizes were 1.82 μm for the graphite and 0.36 to 0.69 μm for the fog-oil. A depression in body weight was found only in animals exposed to the graphite or the graphite and fog-oil mixture. However, the authors did not consider the finding to be toxicologically relevant, mainly because no effects on body weight were seen in a longer study (Aranyi et al. 1992).

In the longer study, rats were exposed for 13 weeks to mixtures of fog oil and graphite (Aranyi et al. 1992). The MMAD of the particles ranged from 0.40 to 0.69 μm. The lowest-concentration exposures were fog oil at 250 mg/m^3 and graphite at 100 mg/m^3 and the highest-concentration exposures were graphite at 200 mg/m^3 and fog oil at 1,000 mg/m^3. No overt signs of toxicity were observed during exposure; particularly, no differences in body weight were noted. Histopathological changes were found in the lungs of all animals exposed to the aerosol mixtures. The main finding was hyperplasia of the goblet cells in the nasal epithelium. Upon removal from the test chambers, the changes were reversible. The epithelia of the terminal bronchioles and of the alveoli showed some hyperplasia. The incidence of those changes was 100% in all exposure groups. In a recovery study, Aranyi et al. (1992) found that the changes did not disappear, presumably because of the continued presence of graphite particles in the lung. In addition, pulmonary and mediastinal lymph nodes and lymphoid tissues showed hyperplasia and graphite-containing granulomas.

Oral Exposures

Graphite type 6353 and Microfyne dissolved in corn oil were not acutely toxic in rats at oral exposures of 5 grams per kilogram (g/kg) of body weight (Manthei et al. 1980).

Intratracheal Exposures

A suspension containing 100 mg of graphite was instilled intratracheally into rats (Ray et al. 1951). Only a minimal pulmonary-tissue reaction was produced; most of the material disappeared from the lung and only a slight reticulosis occurred in the alveolar tissue.

Intraperitoneal Injection

Bovet (1952) injected aqueous suspensions of graphite (5 to 10 mg; particle size, 3 and 10 μm) into the peritoneal cavity of mice. After initial foreign-body inflammatory changes, fibrous nodules developed.

Dermal Exposures

Micro-260 (500 mg suspended in saline) was applied to the clipped skin of rabbits under a gauze pad covered with a polyethylene film (Manthei and Heitkamp 1988). After 24 hr the graphite was removed by gentle washing, and the skin was inspected for signs of irritation. No signs were observed. The test was repeated with graphite dissolved in a slurry of corn oil, and again no signs of skin irritation were observed. The negative findings were confirmed by histopathological examination of the tested skin areas. Negative results were also reported in a study that examined the skin irritation (24 hr) of graphite type 6353 and Microfyne (Manthei et al. 1980). No signs of toxicity in general and of skin irritation in particular were observed with applied doses of 2 g/kg.

Ocular Exposures

Micro-260 was instilled into the conjunctival sack of one eye in rabbits, the contralateral eye serving as control (Manthei and Heitkamp 1988). A volume of 0.1 milliliter (mL) was used, corresponding to 12 mg of the test material. The eyes appeared normal 1 hr later, but there were signs that the rabbits had removed the graphite with their forepaws. Four of six rabbits had mild redness of the nictitating membrane after 24 hr; the irritation disappeared after 72 hr. Fluorescein stain did not reveal abrasion or other lesions of the cornea.

Graphite type 6353 and Microfyne were deposited in quantities of 100 mg into the conjunctival sac in rabbits (Manthei et al. 1980). The two materials were not found to be eye irritants.

Carcinogenic Effects

No data are available on the carcinogenicity of graphite.

Reproductive and Developmental Effects

No data are available on the reproductive and developmental effects of graphite exposure.

In Vitro Tests

Placke and Fisher (1987) evaluated Micro-260, KS-2, graphite fibers, and nickel-coated graphite fibers together with several other dusts in hamster tracheal-organ cultures. Explants were exposed to the dusts for 1 to 3 weeks. The tissues were examined under the light microscope and scored for metaplastic, dysplastic, and hyperplastic lesions in the mucosa. KS-2 elicited a tissue response thought to represent preneoplastic changes. The changes were somewhat more pronounced than those produced by the other forms of graphite, but were not as well-defined as those resulting from exposure to crocidolite asbestos or nickel- (Ni) coated graphite. Ni-coated graphite fibers together with brass dust were the most cytotoxic materials, whereas the toxicity of the other graphite materials was the same as the toxicity of carbon black or glass beads.

The cytotoxicity of graphite was examined in alveolar macrophages harvested from rat lungs (Anderson et al. 1987). For graphite, the 4-hr lethal concentration for 50% of the test animals (LC_{50}) was found to be 428 µg/mL higher than the LC_{50} for titanium dioxide, carbon, aluminum, or slate aluminum (from 115 to 285 µg/mL).

Graphite type 6353 and Microfyne were evaluated in the Ames mutagenicity test, using five tester strains, with and without an activating system (Manthei et al. 1980). No mutagenic activity was observed.

SUMMARY OF TOXICITY DATA

Table 5-1 summarizes the studies of nonlethal effects of inhalation exposure to graphite particles of the type used by the military as an obscurant. The information on the noncarcinogenic effects are summarized here.

In humans chronically exposed to graphite, graphite pneumoconiosis might develop. The condition is characterized by a granulomatous reaction, interstitial fibrosis, and vascular sclerosis. However, these changes are believed to be due to impurities, particularly silica, in the graphite dust that is mined (Gloyne et al. 1949; Harding and Oliver 1949; Hanoa 1983). Acute and subchronic exposure (up to 13 weeks) studies in animals involving inhalation exposure to graphite, with or without the concomitant presence of fog oil, showed only minimal inflammatory reactions in the respiratory tract. The inflammation was fully reversible most of the time. However, if graphite particles persist in the lung, epithelia of the terminal bronchioles and alveoli show signs of hyperplasia, and graphite-containing granulomas are found in lymphoid tissue. In vitro, graphite has minimal cytotoxicity and no mutagenic activity.

PREVIOUS RECOMMENDED EXPOSURE LIMITS

The American Conference of Governmental Industrial Hygienists (ACGIH) has established a Threshold Limit Value time-weighted average

TABLE 5-1 Summary of Nonlethal Effects of Exposure to Graphite

Species	Exposure Frequency and Duration	NOAEL (mg/m^3)	LOAEL (mg/m^3)	End Point and Comments	Reference
Rat, Fischer 344 (Micro-266)	4 hr, 1 d	100	500	Measured number of pmn leukocytes in BAL fluid and alveolar macrophage activation, transient	Anderson et al. 1989
Rat, Sprague-Dawley (type 6353 and Microfyne)	1 hr, 1 d	23,700	—	No signs of toxicity up to 2 wk after exposure	Manthei et al. 1980
Rat, Fischer 344 (Micro-260 and -650)	4 hr/d, 4 d	—	100	Minimal adverse effects (transient biochemical changes in BAL fluid analysis)	Thomson et al. 1988
Rat, Fischer 344 (Micro-260)	2 hr/d, 5 d/wk, 2 wk	—	100	Mild reversible changes in BAL fluid analysis	Thomson et al. 1990
Rat, Fischer 344 (graphite powder Al)	4 hr/d, 4 d/wk, 4 wk	—	200	Decreased body weight	Aranyi et al. 1992
Rat, Fischer 344 (graphite powder Al)	4 hr/d, 4 d/w, 13 wk	—	100 250 (fog oil)	Epithelial hyperplasia in terminal bronchioles and alveoli, not reversible	Aranyi et al. 1992

Abbreviations: NOAEL, no-observed-adverse-effect level; LOAEL, lowest-observed-adverse-effect level; pmn, polymorphonuclear; BAL, bronchoalveolar lavage.

(TLV-TWA) of 2 mg/m^3 for graphite as a respirable dust (all forms except graphite fibers) (ACGIH 1991).

The Occupational Safety and Health Administration (OSHA) established a permissible exposure limit (PEL)-TWA of 3.0 mg/m^3 for natural graphite with the respirable fraction containing less than 1% quartz. The PEL-TWAs for synthetic graphite are 15 mg/m^3 for total particulate material and 5 mg/m^3 for the respirable fraction (U.S. Department of Labor 1997).

The National Institute for Occupational Safety and Health (NIOSH) established a recommended exposure limit (REL)-TWA of 2.5 mg/m^3 for respirable dust from natural graphite (NIOSH 1996). NIOSH has not established a REL-TWA for synthetic graphite.

SUBCOMMITTEE EVALUATION AND RECOMMENDATIONS

MILITARY EXPOSURES

Emergency Exposure Guidance Levels (EEGLs)

No acute toxicity data are available that would indicate adverse health effects caused by graphite in humans. In acute inhalation studies with rats, no deaths occurred even at the highest concentrations used. A 4-hr exposure to synthetic graphite (Micro-260) at 100 mg/m^3 did not produce signs of acute inflammation (Anderson et al. 1989). The extensive inhalation data base for several other poorly soluble, low-toxicity particles (e.g., talc, titanium dioxide, carbon black; see Chapter 4 for a more detailed discussion) is also relevant to recommending emergency exposure limits for graphite. When doses of poorly soluble, low-toxicity particles are >1-2 mg/g of lung in the rat, extensive evidence shows that macrophage-mediated particle-clearance mechanisms are overloaded. That response results in an accumulation of particles in the lung (Morrow et al. 1991; Oberdörster 1994). Associated with the overload of particle clearance by low-toxicity particles in the rat is the development of pulmonary inflammation, fibrosis, and epithelial hyperplasia; inhalation exposures resulting in lung burdens below 1-2 mg/g of lung appear to be without adverse effect (Morrow et al. 1991; Oberdorster 1994). Analysis of the relation between lung-particle dose and particle-clearance overload indicates that clearance overload is more closely related to the particle volume than to the mass of particulate material in the lung

(Morrow 1988; Morrow et al. 1991). In this respect, for graphite, with a density of 2.2 g/cm^3, a lung mass burden of approximately >2 mg/g of lung would be expected to cause particle-clearance overload and adverse lung effects (Morrow et al. 1991).

The approach the subcommittee used to recommend EEGLs for graphite was to estimate the maximal acute exposure concentration resulting in a lung-particle dose below that associated with impairment of alveolar clearance and adverse lung effects. Extrapolating from the data base on long-term inhalation studies using low-toxicity, poorly-soluble particles in rats, a lung dose of graphite at <2 mg/g of lung (incorporates a correction for the density of graphite) would not be expected to alter alveolar particle clearance and produce adverse effects in the rat lung. Because the lung dose associated with overload clearance is derived from long-term inhalation studies, the subcommittee considered the possibility that the response could be greater to a dose delivered by acute exposure (high-dose rate) than by chronic exposures. To address the potential for greater responses due to the high-dose rate in acute inhalation exposures, the subcommittee reduced the dose of 2 mg/g of lung by a factor of 10, and estimated the maximal 15-min, 1-hr, and 6-hr exposure concentrations that would result in a lung dose of <0.2 mg/g of lung. Making reasonable assumptions for minute volume (43 L/min, heavy work activity; Diem and Lentner 1970), deep-lung particle deposition (35% of respirable particles; Stahlhofen et al. 1989) and lung weight (1,000 g; Diem and Lentner 1970), the subcommittee calculated that a 15-min exposure to graphite at approximately 880 mg/m^3 is the maximal concentration resulting in a deep-lung dose of <0.2 mg/g of lung that would not cause significant adverse effects.[1] Support for that calculation can be found in the acute inhalation study in which exposure of rats to graphite at either 500 mg/m^3 for 4 hr or 100 mg/m^3 for 4 hr per day for 4 days resulted in only transient pulmonary inflammation (Anderson et al. 1989; Thomson et al. 1986). It is estimated that the deep-lung particle burden in the rat studies was approximately 2-2.5 mg/g of lung (assumes a rat minute volume of 200 mL/min and a particle deposition

---

[1] Calculation of 15-min exposure concentration in humans resulting in a lung graphite dose of 0.2 mg/g of lung.

Exposure concentration = [(0.2 mg/g of lung)(1,000 g/lung)]/
　　　　　　　　　　　[(43 L/min)(15 min)(m^3/1,000)(0.35)].

of 10%), which is just at or above the dose associated with clearance overload in rats. Also supporting a 15-min EEGL of approximately 880 mg/m^3 is the acute inhalation study of Manthei et al. (1980), who reported that a 1-hr exposure of rats to graphite at 23,700 mg/m^3 did not produce any gross signs of toxicity. Extrapolating from the 15-min EEGL, a 1-hr EEGL of 220 mg/m^3 and a 6-hr EEGL of 40 mg/m^3 (rounded from 36.6 mg/m^3) are recommended.

Repeated Exposure Guidance Level (REGL)

Because the graphite used by the Army is free of silica and has properties similar to other low-toxicity, poorly soluble particles (e.g., titanium dioxide, talc, and carbon black), the subcommittee believes it is reasonable to recommend a REGL by determining the maximal long-term exposure concentration that would not result in a lung burden that would overload particle-clearance mechanisms (after accounting for the density of graphite). The scientific rationale for this approach is discussed above in the context of recommending EEGLs. Making reasonable assumptions for minute ventilation (22 L/min, moderate work activity; Raabe 1979), the proportion of respirable particles depositing in the deep lung (35%; Stahlhofen et al. 1989), lung weight (1,000 g; Diem and Lentner 1970), and particle-clearance half-time (500 days; Bohning et al. 1982) for humans, the subcommittee estimated that chronic exposure to graphite at 1 mg/m^3 would result in a lung burden of <2 mg/g of lung.[2] The calculation takes into account sources of uncertainty in lung dosimetry between rodents and humans and, therefore, an additional uncertainty factor extrapolating from animals to humans was not applied. Additionally, the uncertainty factor was reduced to 1 because the rat is a more sensitive species than the human to the effects of poorly soluble particles in the lung. The subcommittee recommends a REGL for graphite of 1 mg/m^3 for 8 hr per day, 5 days per week.

---

[2]Calculation of exposure concentration (8 hr per day × 5 days per week) resulting in a lung dose of 2 mg/g of lung.

$$\text{Exposure concentration} = [(2 \text{ mg/g lung})] \div \{[(22 \text{ L/min})(60 \text{ min})(8 \text{ h/d})(5 \text{ d/7 d})(\text{m}^3/1{,}000)(0.35)(\text{lung}/1{,}000 \text{ g})] \div [\ln 2 / 500 \text{ d}]\}$$

Public Exposures

Short-Term Public Emergency Guidance Levels (SPEGL)

The general public may be assumed to have a greater range of individual susceptibilities compared with healthy military personnel. The subcommittee applied an uncertainty factor of 10 to take that into account. Accordingly, the 15-min SPEGL is 88 mg/m^3, the 1-hr SPEGL is 22 mg/m^3, and the 6-hr SPEGL is 4.0 mg/m^3.

Repeated Public Exposure Guidance Level (RPEGL)

The subcommittee recommends that the RPEGL be derived by incorporating an uncertainty factor of 10 into the REGL to account for the potentially greater range of susceptibilities in the general public compared with healthy military personnel. Accordingly, the RPEGL for graphite is 0.1 mg/m^3.

Summary of Subcommittee Recommendations

Table 5-2 summarizes the subcommittee's recommendations for the EEGLs and the REGL for military personnel, and Table 5-3 summarizes the recommendations for the SPEGLs and the RPEGL at the boundary of military facilities.

RESEARCH NEEDS

According to available data, graphite dusts behaves biologically like dusts with low biological activity both in vivo and in vitro. Those dusts do not produce signs of acute toxicity and might produce signs of chronic lung disease only under conditions of overload (see Chapter 4). The recommendations for EEGLs were derived by calculating exposure concentrations that would not yield sufficient deposited graphite to cause lung overload. The calculations were verified against biological data on rats for which the calculated graphite lung doses were such that conditions of lung overload were expected. In those cases, biological effects, notably inflammation, were observed.

Because the biological data forming the basis for that verification are

TABLE 5-2 EEGLs and REGL for Graphite Smoke for Military Personnel

Exposure Guideline	Exposure Duration	Exposure Guidance Level (mg/m³)
EEGL	15 min	880
	1 hr	220
	6 hr	40
REGL	8 hr/d, 5 d/wk	1

Abbreviations: EEGL, emergency exposure guidance level; REGL, repeated exposure guidance level.

TABLE 5-3 SPEGLs and RPEGL for Graphite Smoke for the Boundaries at Military-Training Facilities

Exposure Guideline	Exposure Duration	Exposure Guidance Level (mg/m³)
SPEGL	15 min	88
	1 hr	22
	6 hr	4
RPEGL	8 hr/d, 5 d/wk	0.1

Abbreviations: SPEGL, short-term public emergency guidance level; RPEGL, repeated public exposure guidance level.

sparse, the subcommittee recommends that an acute inhalation study of graphite in rats be conducted, uising multiple concentrations to describe fully the exposure dose-response relationship. Concentrations expected to yield lung burdens of graphite at or below the concentration expected to result in lung overload could be used. End points monitored should include lung burdens of graphite, histopathological, changes and markers of inflammation.

REFERENCES

ACGIH (American Conference of Governmental Industrial Hygienists). 1991. Documentation of the Threshold Limit Values and Biological Exposure Indices, 6th Ed. American Conference of Governmental Industrial Hygienists, Cincinnati, Ohio.

Anderson, R.S., S.M. Thomson, and L.L. Gutshall. 1989. Comparative effects of inhaled silica or synthetic graphite dusts on rat alveolar cells. Arch. Environ. Contam. Toxicol. 18:844-849.

Anderson, R.S., L.L. Gutshall, and S.A. Thomson. 1987. Rat Pulmonary Alveolar Macrophages In Vitro Cytotoxicity to Six Metal Dusts. CRDEC-TR-88037. Chemical Research, Development and Engineering Center, U.S. Army Armament, Munitions and Chemical Command, Aberdeen Proving Ground, Edgewood, Md.

Aranyi, C., N. Rajendran, and J. Bradof. 1992. Thirteen-week Inhalation Toxicity Study with Aerosol Mixtures of Fog Oil and Graphite Particles in F344/N Male Rats. DAMD17-89-C-9043. Prepared by IIT Research Institute, Chicago, for U.S. Army Medical Research and Development Command, Fort Detrick, Md.

Bohning, D.E., H.L. Atkins, and S.H. Cohn. 1982. Long-term particle clearance in man: Normal and impaired. Ann. Occup. Hyg. 26:259-271.

Bovet, P. 1952. Die wirkung von graphit und anderen kohlenstoffmodifikationen im tierversuch; zugleich ein beitrag zur experimentellen silikoseforschung. Schweiz. Z. Allg. Pathol. Bakteriol. 15:548-565.

Driver, C.J., M.W. Ligotke, W.G. Landis, J.L. Downs, B.L. Tiller, and E.B. Moore. 1993. Environmental and Health Effects Review for Obscurant Graphite Flakes. ERDEC-TR-056. Edgewood Research, Development and Engineering Center, U.S. Army Armament, Munitions and Chemical Command, Aberdeen Proving Ground, Edgewood, Md.

Dunner, L., and J.T. Bagnall. 1946. Graphite pneumoconiosis complicated by cavitation due to necrosis. Br. J. Radiol. 19(220):165-168.

Gaensler, E.A., J.B. Cadigan, A.A. Sasahara, E.O. Fox, and H.E. MacMahon. 1966. Graphite pneumoconiosis of electrotypers. Am. J. Med. 41:864-882.

Gloyne, R.S., G. Marshall, and C. Hoyle. 1949. Pneumoconiosis due to graphite dust. Thorax 4:31-38.

Diem, K., and C. Lentner, eds. 1970. Documenta Geigy: Scientific Tables, 7th Ed. Basle, Switzerland: Ciba-Geigy.

Guelta, M.A., D.R. Banks, and R.S. Grieb. 1993. Particle Size Analysis of Graphite Obscurant Material from an XM56 Smoke Generator. ERDEC-TR-053. Edgewood Research, Development and Engineering Center, U.S. Army Armament, Munitions and Chemical Command, Aberdeen Proving Ground, Edgewood, Md.

Hanoa, R. 1983. Graphite pneumoconiosis. A review of etiologic and epidemiologic aspects. Scand. J. Work Environ. Health 9:303-314.

Harding H.E., and G.B. Oliver. 1949. Changes in the lungs produced by natural graphite. Br. J. Ind. Med. 6:91-99.

Jaffé, F.A. 1951. Graphite pneumoconiosis. Am. J. Pathol. 17:909-923.

Lister, W.B., and D. Wimborne. 1972. Carbon pneumoconiosis in a synthetic graphite worker. Br. J. Ind. Med. 29:108-110.

Lundy, D., and J. Eaton. 1994. Occupational Health Hazards Posed by Inventory U.S. Army Smoke/Obscurant Munitions (Review Update). WRAIR/RT-

94-0001. AD-A276 774. Prepared by U.S. Army Medical Research Detachment, Wright-Patterson Air Force Base, Ohio, for Walter Reed Army Institute of Research, Washington, D.C.

Manthei, J.H., and D.H. Heitkamp. 1988. Irritant Hazard Assessment of Graphite "Micro 260" in the Rabbit. CRDEC-TR-88077. Chemical Research, Development and Engineering Center, U.S. Army Armament, Munitions and Chemical Command, Aberdeen Proving Ground, Edgewood, Md.

Manthei, J.H., F.K. Lee Jr., M. Donnelly, and J.T. Weimer. 1980. Preliminary Toxicity Screening Studies of 11 Smoke Candidate Compounds. ARCSL-TR-79056. Chemical Systems Laboratory, U.S. Army Armament, Munitions and Chemical Command, Aberdeen Proving Ground, Edgewood, Md.

Mazzucchelli, L., H. Radelfinger, and R. Kraft. 1996. Nonasbestos ferruginous bodies in sputum from a patient with graphite pneumoconiosis: A case report. Acta Cytol. 40:552-554.

Morrow, P.E. 1988. Possible mechnanisms to explain dust overloading of the lungs. Fundam. Appl. Toxicol. 10:369-384.

Morrow, P.E. 1992. Dust overloading of the lungs: Update and appraisal. Toxicol. Appl. Pharmacol. 113:1-12.

Morrow, P.E., H. Muhle, and R. Mermelstein. 1991. Chronic inhalation study findings as a basis for proposing a new occupational dust exposure limit. J. Am. Coll. Toxicol. 10:279-290.

NIOSH (National Institute for Occupational Safety and Health). 1996. Graphite (natural). In Documentation for Immediately Dangerous to Life or Health. Online. Entry last updated Aug. 16, 1996. Available: http://www.cdc.gov/niosh/idlh/13463677.html. U.S. Department of Health and Human Services, National Institute for Occupational Safety and Health, Division of Standards Development and Technology Transfer, Cincinnati, Ohio.

Oberdörster, G. 1994. Extrapolation of results from animal inhalation studies with particles to humans. Pp. 57-73 in Toxic and Carcinogenic Effects of Solid Particles in the Respiratory Tract, D.L. Dungworth, J.L. Mauderly, and G. Oberdörster, eds. Washington, D.C.: ILSI Press.

Pendergrass, E.P., A.J. Vorwald, M.M. Mishkin, J.G. Whildin, and C.W. Werley. 1968. Observations on workers in the graphite industry. II. Med. Radiogr. Photogr. 44:2-17.

Petsonk, E.L., E. Storey, P.E. Becker, C.A. Davidson, K. Kennedy, V. Vallyathan. 1988. Pneumoconiosis in carbon electrode workers. J. Occup. Med. 30:887-891.

Placke, M.E., and G.L. Fisher. 1987. In Vitro Toxicity Evaluation of Ten Particulate Materials in Tracheal Organ Culture. CRDEC-TR-88010. Chemical Research, Development and Engineering Center, U.S. Army Armament, Munitions and Chemical Command, Aberdeen Proving Ground, Edgewood, Md.

Raabe, O.G. 1979. Deposition and Clearance of Inhaled Aerosols. UCD-472-503. Laboratory for Energy-Related Health Research, University of California at Davis for the U.S. Department of Energy, Washington, D.C. Available from NTIS, Springfield, Va., Doc. No. NTIS/UDC-472-543.

Ranasinha, K.W., and C.G. Uragoda. 1972. Graphite pneumoconiosis. Br. J. Ind. Med. 29:178-183.

Ray, S.C., E.J. King, and C.V. Harrison. 1951. The action of small amounts of quartz and larger amounts of coal and graphite on the lungs of rats. Br. J. Ind. Med. 8:68-73.

Stahlhofen, W., G. Rudolf, and A.C. James. 1989. Intercomparison of experimental regional aerosol deposition data. J. Aerosol Med. 2(3):285-308.

Taylor, H.A., Jr. 1985. Graphite. Pp. 339-348 in Mineral Facts and Problems. U.S. Bureau of Mines Bulletin 675. Washington, D.C.: U.S. Government Printing Office.

Thomson, S.A., D.C. Burnett, R.J. Hilaski, J.T. James, and W.G. Landis. 1986. Environmental and Inhalation Hazards from Repeated Exposure to EA 5768. Pp. 111-126 in Proceedings of Smoke/Obscurants Symposium X. AMCPM-SMK-T-003-86. Chemical Research, Development and Engineering Center, U.S. Army Armament, Munitions and Chemical Command, Aberdeen Proving Ground, Edgewood, Md.

Thomson, S.A., D.W. Johnson, R.S. Anderson, and D.C. Burnett. 1987. Environmental and Inhalation Hazards from Acute Exposure to EA 5768. Pp. 53-68 in Proceedings of Smoke/Obscurants Symposium XI. AMCPM-SMK-T-002-87. Chemical Research, Development and Engineering Center, U.S. Army Armament, Munitions and Chemical Command, Aberdeen Proving Ground, Edgewood, Md.

Thomson, S.A., J.D. Bergmann, D.C. Burnett, J.C. Carpin, C.L. Crouse, R.J. Hilaski, B. Infiesto, and E. Lawrence-Beckett. 1988. Comparative Inhalation Screen of Titanium Dioxide and Graphite Dusts. CRDEC-TR-88161. Chemical Research, Development and Engineering Center, U.S. Army Armament, Munitions and Chemical Command, Aberdeen Proving Ground, Edgewood, Md.

Uragoda, C.G. 1989. Graphite pneumoconiosis and its declining prevalence in Sri Lanka. J. Trop. Med. Hyg. 92:422-424.

U.S. Department of Labor. 1997. Occupational Safety and Health Standards. Air Contaminants. Code of Federal Regulations, Title 29, Part 1910, Section 1910.1000. Washington, D.C.: U.S. Government Printing Office.

Watson, A.J., J. Black, A.T. Doig, and G. Nagelschmidt. 1959. Pneumoconiosis in carbon electrode makers. Br. J. Ind. Med. 16:274-285.

Zahorski, W. 1960. Pneumoconiosis in workers of artificial graphite plants [in Polish]. Med. Pr. 11:383-386.

Zahorski, W., Z. Potoska-Skowronek, and W. Pierzchala. 1975. Pneumoconiosis in persons engaged in the production of carbon electrodes [in Polish]. Med. Pr. 26:1-8.